PERSPECTIVES ON CONTEMPORARY STATISTICS

David C. Hoaglin
David S. Moore
Editors

The Mathematical Association of America

MAA Notes and Reports Series

The MAA Notes and Reports Series, started in 1982, addresses a broad range of topics and themes of interest to all who are involved with undergraduate mathematics. The volumes in this series are readable, informative, and useful, and help the mathematical community keep up with developments of importance to mathematics.

MAA Notes

1. Problem Solving in the Mathematics Curriculum,
    Committee on the Teaching of Undergraduate Mathematics,
    a subcommittee of the Committee on the Undergraduate Program in Mathematics, *Alan H. Schoenfeld,* Editor.

2. Recommendations on the Mathematical Preparation of Teachers,
    Committee on the Undergraduate Program in Mathematics, Panel on Teacher Training.

3. Undergraduate Mathematics Education in the People's Republic of China,
    Lynn A. Steen, Editor.

4. Notes on Primality Testing and Factoring,
    Carl Pomerance.

5. American Perspectives on the Fifth International Congress on Mathematical Education,
    Warren Page, Editor.

6. Toward a Lean and Lively Calculus,
    Ronald G. Douglas, Editor.

7. Undergraduate Programs in the Mathematical and Computer Sciences: 1985–86,
    D. J. Albers, R. D. Anderson, D. O. Loftsgaarden, Editors.

8. Calculus for a New Century,
    Lynn A. Steen, Editor.

9. Computers and Mathematics: The Use of Computers in Undergraduate Instruction,
    Committee on Computers in Mathematics Education, D. A. Smith, G. J. Porter, L. C. Leinbach, and R. H. Wenger,
    Editors.

10. Guidelines for the Continuing Mathematical Education of Teachers,
    Committee on the Mathematical Education of Teachers.

11. Keys to Improved Instruction by Teaching Assistants and Part-Time Instructors,
    Committee on Teaching Assistants and Part-Time Instructors, Bettye Anne Case, Editor.

12. The Use of Calculators in the Standardized Testing of Mathematics,
    John Kenelly, Editor, published jointly with The College Board.

13. Reshaping College Mathematics,
    Committee on the Undergraduate Program in Mathematics, Lynn A. Steen, Editor.

14. Mathematical Writing,
    by *Donald E. Knuth, Tracy Larrabee, and Paul M. Roberts.*

15. Discrete Mathematics in the First Two Years,
    Anthony Ralston, Editor.

16. Using Writing to Teach Mathematics,
    Andrew Sterrett, Editor.

17. Priming the Calculus Pump: Innovations and Resources,
    Committee on Calculus Reform and the First Two Years,
    a subcommittee of the Committee on the Undergraduate Program in Mathematics, *Thomas W. Tucker,* Editor.

MAA Reports

First Printing
© 1992 by the Mathematical Association of America
ISBN 0-88385-075-3
Library of Congress Catalog
Card Number 91-062170
Printed in the United States of America

PERSPECTIVES ON CONTEMPORARY STATISTICS

CONTENTS

Chapter 5 THE STATISTICAL APPROACH TO DESIGN
OF EXPERIMENTS 71

Ronald D. Snee and Lynne B. Hare

Chapter 6 WHAT IS PROBABILITY? 93

Glenn Shafer

Chapter 7 THE REASONING OF STATISTICAL INFERENCE . . . 107

Lincoln E. Moses

Chapter 8 DIAGNOSTICS 123

David C. Hoaglin

Chapter 9 RESISTANT AND ROBUST PROCEDURES 145

Thomas P. Hettmansperger and Simon J. Sheather

Preface

This volume concerns, to paraphrase a famous title of Felix Klein, elementary statistics from an advanced standpoint. But unlike Klein, who presented a technically advanced discussion of elementary topics, we focus on topics central to the teaching of statistics to beginners and offer expositions that are guided by the current state of statistical research and practice.

All scientific fields face the problem of keeping instruction in touch with the fruits of research. Because statistics is a methodological discipline that is widely applied, the influence of research on beginning instruction is mediated through its influence on statistical practice. We have asked our contributors to let good current practice guide them. Statistical practice has changed radically during the past generation under the impact of ever cheaper and more accessible computing power. Beginning instruction has lagged behind the evolution of the field. Software now enables students to shortcut unpleasant calculations, but this is only the most obvious consequence of changing statistical practice. The content and emphases of statistics instruction still need much rethinking.

Changing instruction in statistics presents special challenges because a wide range of departments and programs offer courses on the subject, often taught by instructors who are not primarily trained as statisticians. We are eager to minimize the time required for changes in statistical practice, especially those driven by advances in computing, to affect instruction in introductory courses. Thus our intended audience includes mathematicians, engineers, and social scientists who teach beginning statistics, as well as graduate students and others new to the field.

This volume assembles nine new essays on important topics in present-day statistics, selected because we believe they should influence the teaching of statistics at the college level and elsewhere. Students approach statistics from diverse disciplinary backgrounds and with various levels of mathematical preparation. Accordingly, most chapters present modern perspectives on central aspects of statistics and emphasize the conceptual content that should accompany all varieties of beginning instruction; Chapters 8 and 9 offer expositions of more advanced topics that should influence teaching but will appear in some detail only in courses for well-prepared students.

As David Moore discusses more fully in Chapter 1, it is useful to divide statistics, as a science, into three broad areas:

- Analyzing data.

- Producing data.

- Making inferences from data.

Chapter 1 then builds on this framework to provide a contemporary overview of statistics as the science of data—a view much broader than the "inference from data," emphasized by much traditional teaching. Data analysis has emerged as a major area in both research and practice. In Chapter 2, Paul Velleman and David Hoaglin discuss the philosophy and some of the tools of data analysis.

Both in data analysis and inference, computing has had a profound impact on contemporary statistics. Ronald Thisted and Paul Velleman describe this impact, and its implications for teaching, in Chapter 3.

Firm conclusions usually require careful design for production of appropriate data—the most secure setting for formal statistical inference. In Chapter 4, Judith Tanur discusses the science of survey sampling, and in Chapter 5 Ronald Snee and Lynne Hare present essential concepts of statistical design of experimentation.

Formal statistical inference remains a very important part of statistical practice, and it uses mathematical models that combine nonrandom and probabilistic components. This role for probability implies that alternative interpretations of the mathematics of probability will lead to alternative approaches to statistical inference. In Chapter 6, Glenn Shafer describes the variety of contemporary ideas of probability, and in Chapter 7 Lincoln Moses discusses the reasoning of formal inference.

Statistical practice involves a dialogue between data and models, in which the data can criticize and even falsify the model. In many situations, diagnostic tools are available to reveal a variety of difficulties. An alternative approach abandons the more traditional inference methods for others that are less sensitive to unusual data points and to violations of the model assumptions. In Chapter 8, David Hoaglin gives an introduction to diagnostics, primarily in the important setting of regression models. In Chapter 9, Thomas Hettmansperger and Simon Sheather discuss the alternative approach of resistant and robust procedures.

Returning to our main aim of providing input for introductory teaching, we group the chapters as follows. Chapters 2 to 5 exposit areas that are important in practice but are underrepresented in traditional instruction and hence deserve more attention. Chapters 6 and 7 consider the heart of the traditional content, probability and inference, from a conceptual point of view influenced by contemporary issues. Chapters 8 and 9 introduce newer topics that have become standard in practice under the influence of fast and easy computing.

In developing the chapters the authors have drawn on their expertise in the topics and on their experience in both teaching and practice. We promised them considerable flexibility, and we encouraged them:

- To be only as technical as necessary.

- To use examples freely.

- To make general remarks and philosophical comments that would help readers appreciate the intent and spirit of statistics.

- To comment on priorities and pitfalls in teaching.

- To give references for further reading, especially to real applications and to more detailed and technical exposition.

They have given us chapters that generously fulfill these goals.

As editors we appreciate that others, viewing the science of statistics from different perspectives, might have selected a somewhat different set of topics or might have struck a different balance. Thus we have not aimed for comprehensive coverage of our changing field. Instead, we are pleased to offer a selection of lively materials that should help instructors in a variety of courses give their students a stimulating appreciation of contemporary statistical practice.

Acknowledgments

We are grateful to many people for their contributions to this volume. Warren Page, editor of the *MAA Notes* series, invited us to tackle the project. Lynne B. Hare, Thomas P. Hettmansperger, Lincoln E. Moses, Glenn Shafer, Simon J. Sheather, Ronald D. Snee, Judith M. Tanur, Ronald A. Thisted, and Paul F. Velleman, the authors of the chapters, readily agreed to participate and cheerfully cooperated through drafts, revisions, and editing. Staff members of the Department of Statistics at Purdue University helped with a number of aspects of production. The publications staff of the MAA gave us valuable advice and efficiently processed the completed manuscript. The National Science Foundation provided partial support for David Moore's work under grant DMS-8901922 and for David Hoaglin's work under grant SES-8908641.

David C. Hoaglin
David S. Moore

Cambridge, Massachusetts
West Lafayette, Indiana
October 1991

Introduction: What Is Statistics?

David S. Moore
Purdue University

1. THE SCIENCE OF DATA

Statistics is the science of gaining information from data. Data are of course numbers, but they are more than that. Data are *numbers with a context*. The number 10.2, for example, carries no information without a context. But if we hear that a friend's new baby weighed 10.2 pounds at birth, we at once appreciate the healthy size of the child. The context engages our background knowledge and allows us to make judgments. (Contrast our response to a claimed birth weight of 10.2 kilograms or 10.2 ounces.) The number has become informative.

Conclusions based on data seem to be (and often are) firmer than the results of anecdote or speculation. Arguments from data have become ever more common in a growing number of professions and in public policy. Here are some examples of statistical questions raised in the mind of an inquisitive viewer of the nightly news.

- The Bureau of Labor Statistics reports that the unemployment rate last month was 6.5%. How did the government obtain this information? (I wasn't asked if I was employed last month.) How accurate is that 6.5%?

- Another news item mentions restrictions on smoking in public places. I hear that much of the evidence that links smoking to lung cancer and other health problems is "statistical." What kind of evidence is this?

- A medical reporter claims that an experiment has shown that taking aspirin regularly reduces the risk of a heart attack. How can an experiment be designed that convincingly shows this?

- Here's a special report on the international competitiveness of American industry. The experts interviewed talk about improving quality and productivity through better management, new technology, and effective use of statistics. What can statistics do here except keep score?

These questions are strikingly different from the issues we most often address in teaching statistics. A mathematics student who takes a typical course on probability and statistics is ill-prepared to answer such questions. Students in traditional statistical methods courses are only a bit better off. A wide gap separates statistics teaching from statistical practice. This volume presents a view of statistics that reflects statistical practice as well as changing emphases in statistics research and in the approach that statisticians take to their subject.

A wide gap separates statistics teaching from statistical practice. This volume presents a view of statistics that reflects statistical practice as well as changing emphases in statistics research and in the approach that statisticians take to their subject.

To appreciate some of the distinctions between traditional instruction and contemporary practice, it is helpful to divide the science of data into three broad areas:

- Analyzing data.

- Producing data.

- Inference from data.

Statistics teaching has traditionally concentrated on the third area, and has understood "inference from data" to mean *formal inference* based on probability theory. Analyzing data received brief coverage under the heading "descriptive statistics," and statistical designs for producing data were largely relegated to specialist training. Contemporary practice and research have quite different emphases. Although formal inference remains important, we increasingly recognize that its scope is too restricted to regard inference as the heart of statistics. *Data analysis*, which provides tools and strategies for extracting information from data, not only as a preliminary to formal inference but also in settings where formal inference is not justified, has emerged as a major area in both research and practice. Paul Velleman and David Hoaglin discuss the philosophy and some of the tools of data analysis in Chapter 2. The emergence of data analysis owes much to ever faster and more convenient computing. Software implements both calculation and graphics and increasingly helps as well to guide the user. The impact of computing in contemporary statistics has been profound, not only in data analysis but in inference. Ronald Thisted and Paul Velleman describe this impact, and its implications for teaching, in Chapter 3.

We recommend that instruction in modern statistics begin with data analysis, both because concrete experience with data motivates the more abstract parts of our subject and because exploring even haphazardly produced data can provide insight. But firm conclusions usually require careful production of appropriate data. Basic concepts for *designing data production* (probability sampling, randomized comparative experiments) are arguably the most influential of all statistical ideas. The gradual adoption of these ideas in many areas of applied science, beginning in the 1920s, has greatly increased the use of data and the quality of the data used. In addition, careful design of data production provides the most secure setting for formal statistical inference. In Chapter 4, Judith Tanur discusses the science of survey sampling, with emphasis on surveys of large dispersed human populations. Ronald Snee and Lynne Hare present essential concepts of statistical design of experimentation in Chapter 5.

Formal inference remains, of course, a very important part of statistical practice, and it remains anchored in mathematical analysis. We can often use inference procedures known to be optimal in some appropriate sense,

and we can accompany conclusions by a statement expressing our confidence in their correctness. Both optimality of inference procedures and statements of the correctness of their outcomes are expressed in the language of *probability*. Varied interpretations of the mathematics of probability give rise to varied approaches to statistical inference. Glenn Shafer describes the variety of contemporary ideas of probability in Chapter 6, followed by Lincoln Moses' discussion of the reasoning of formal inference in Chapter 7.

Formal statistical inference is based on a mathematical model for the data, combining both nonrandom and probabilistic components. Statistical theory deduces the properties of inference procedures from the model. Statistical practice, however, is a dialogue between data and models. The data are invited to criticize and even falsify the model. Instruction that emphasizes only one side of this dialogue (the use of model-based procedures to analyze data) misses an essential part of practice. The other direction of the dialogue is provided by *diagnostic tools* that assess the adequacy of models. The development of new diagnostic procedures is a major area of statistical research. David Hoaglin devotes Chapter 8 to diagnostics, primarily in the important setting of regression models.

We recommend that instruction in modern statistics begin with data analysis, both because concrete experience with data motivates the more abstract parts of our subject and because exploring even haphazardly produced data can provide insight.

What ought the working statistician do when diagnostic work reveals that standard methods (based, for example, on least-squares fitting of linear models with normal variation) are not appropriate? In some circumstances, removing compromised observations and transforming the data to a different scale can eliminate the difficulty. An alternative approach abandons the more traditional inference methods for others that are less sensitive to unusual data points and to violations of the model assumptions that justify the traditional methods. Such *resistant and robust* procedures are another active research area. Thomas Hettmansperger and Simon Sheather conclude the volume with a discussion of this area in Chapter 9.

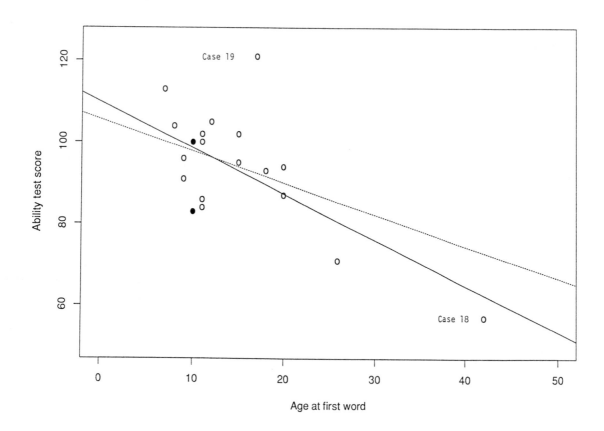

Figure 1: An influential observation: The solid line is the least-squares fit for all observations, the dashed line is the least-squares fit with case 18 removed.

The chapters in this volume address major areas of modern statistics with an emphasis on ideas and practice. In this introductory chapter I attempt an overview of the discipline, comment on the relations between the topics of the later chapters, and describe the relationship between statistics and mathematics.

2. DATA ANALYSIS

Data analysis is the oldest aspect of statistics. Many methods central to statistical practice, such as least squares, originated in data analysis and only later acquired a probabilistic justification. After a period of neglect, data analysis has been reborn, with new methods, greater emphasis on graphics, and a consistent philosophy due to John Tukey. (Volumes 3 and 4 of Tukey's *Collected Works* contain many of his writings in this area. See Jones 1986.)

The essence of data analysis is to "let the data speak," to look for patterns in data without at first considering whether the data are representative of some larger universe. Inspection of data often uncovers unexpected features. If the data were produced to answer a specific question—this is the setting in which formal inference procedures such as confidence intervals and significance tests are best justified—the unusual features may lead us to reconsider the analysis we had planned. Careful data analysis therefore precedes formal inference in good statistical practice. In other situations, we do not have specific questions in mind and want to allow the data to suggest conclusions that we can seek to confirm by further study. We then speak of "exploratory data analysis," on the analogy of an explorer entering unknown lands. Here is an simple example.

Example. Figure 1 presents a scatterplot of data on the age in months at which each of a group of 21 children spoke their first word and their later scores on a test of mental ability. The solid dots each represent two children with identical data. Does age at first word help us predict the later test score? (These data were originally produced by L. M. Linde and published in Mickey, Dunn, and Clark 1967; they have often served as an example in the literature on data analysis and diagnostics.)

A "routine" analysis might simply calculate the correlation coefficient, $r = -0.640$, and the least-squares regression line for predicting test score from age at first word, the solid line in the figure. But the first commandment of data analysis is to *look at the data*. The scatterplot shows a weak linear relationship with two outlying observations, labeled as cases 18 and 19 in the plot. Statistical software allows immediate investigation, with results displayed graphically as well as numerically. Case 19, although far from the regression line, does not have a large influence on the position of the line or the value of the correlation r. Case 18, on the other hand, is highly *influential*. Removing this point moves the regression line to the dashed line in Figure 1 and reduces the correlation to $r = -0.335$. The evidence that age at first word predicts later ability scores becomes much weaker without case 18.

This example illustrates several points that later chapters will elaborate. First, data analysis is not merely a collection of clever graphics such as stem-and-leaf displays and boxplots. It offers general strategies for investigating data. Our example does not use any innovative tools. It does employ two basic strategies: (1) Start with graphic display and move to numerical summaries. Here the scatterplot displays the data, and correlation and the fitted line summarize specific aspects. (2) Look first for an overall pattern (weakly linear here) and then for striking deviations from that pattern (the outlying data points). The example also illustrates why data analysis is an essential prerequisite to formal inference. The t statistic for the significance of the regression of test score on age at first word is $t = -3.63$ when all data points are included, and $t = -1.51$ when case 18 is omitted. The significant relationship suggested by routine analysis of the full data set is suspect, because it depends so strongly on a single observation.

A closely related point is the great sensitivity of standard methods such as least-squares regression to a few influential observations. Current statistical practice offers two general strategies for dealing with this problem. We can attempt to detect the discordant observations, as-

sess their influence on our conclusions, and then remove them if there is reason to do so. Child 18, for example, was so slow to speak that the child development expert who owns the data might consider that this individual does not belong to the population under investigation. If this child is to be included in the target population, more observations are needed to fill the gap in age at first word between case 18 and the other children, reducing the undue influence of this single data point. It is much harder to detect unusual observations when many variables are present, but the diagnostic tools described in Chapter 8 usually succeed when used with care.

An alternative approach to the problem posed by a few strongly influential observations is to employ procedures that are less sensitive to such observations than the usual means, standard deviations, and least-squares fits. Statisticians call such procedures *resistant*. Formal inference based on resistant numerical descriptions may also be *robust*, that is, less sensitive to departures from standard assumptions such as normally distributed errors. Figure 2 contrasts the least-squares regression line for our example (solid) with a regression line fit to the full data set using a resistant procedure (dashed). Many statistical software packages allow a choice between least-squares and resistant procedures; the resistant line in Figure 2 was produced by the Minitab package, a system widely used in elementary instruction. (For a description of the method, see Minitab, Inc. 1989, page 14–11.)

Finally, the example and the developments just mentioned emphasize the importance of computing. We assessed the influence of case 18 by repeating the analysis with this observation omitted, a tedious job by hand even for this small set of two-variable data. "Leave-one-out" calculations are a feature of some common diagnostic methods for high-dimensional data. Although clever algorithms can speed the work, most diagnostic methods are not practical without automated calculation. The same is true of resistant alternatives to least-squares fitting. Least-squares fitting of a linear model requires only the solution of a set of simultaneous linear equations. Resistant methods generally require iterative calculation, with no closed-form solution available.

3. PRODUCING DATA

A well-known mathematical statistician once remarked to me that the evidence linking smoking to lung cancer was "about as strong as statistical evidence can be." This evidence is in fact only about as strong as *non-*

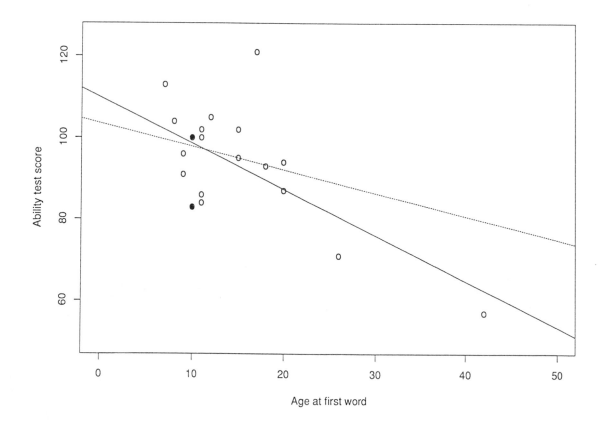

Figure 2: Least-squares fit (solid line) compared with a resistant fitting method (dashed line) applied to the same data.

experimental statistical evidence can be. Contrast the clearer understanding of the director of the Physicians' Health Study, a randomized, comparative, double-blind experiment that demonstrated that taking aspirin regularly reduces the risk of heart attacks in middle-aged men. Shortly thereafter, another study observed several thousand residents of California retirement communities and found that men who chose to take aspirin had no lower risk of heart attacks. The physician who directed the Physicians' Health Study replied (*New York Times*, November 18, 1989) that the authors had "gone beyond the ability of the kind of study they did to draw conclusions of cause and effect" and went on to describe the advantages of controlled experiments over observational studies. Students ought to gain from their first course in statistics a similar understanding of the importance of how data are produced.

Not all data are equally trustworthy. The design of the data-production process is the most important indicator of quality in data. Good data are as much a product

of human skill as hybrid corn and compact disc players. Statistics has much to say about the design of data production. The central contribution is the replacement of human judgment by impersonal chance.

Sampling

Selecting a *sample* to represent some larger population is a common data-production task, one that lies behind opinion polls, government economic statistics, and much social science research. Human choice, either by the selector or by respondents who volunteer their voices, often produces samples that systematically misrepresent the population.

> **Example.** The advice columnist Ann Landers once asked her readers, "If you had it to do over again, would you have children?" Over 10,000 readers wrote in; 70% said "No," and many described the misery their children had brought them. Voluntary response of this kind

overrepresents people with strong feelings, especially strong negative feelings. A national random sample on the same issue found that over 90% of parents *would* choose to have children again. Voluntary response can easily produce 70% "No" in a sample when the truth about the population is close to 90% "Yes." Yet the media continue to conduct voluntary response polls and discuss them as if they provided useful information.

The statistician's remedy is to allow impersonal chance to choose the sample. Selecting a sample at random is obviously fair in the sense of eliminating unintended favoritism. That is, random samples eliminate the kind of bias introduced by voluntary response. Equally important, random sampling guarantees that the mathematics of probability theory applies to the results of sampling. Repeated random sampling from the same population would produce a distribution of outcomes that is centered at the population truth and has a spread determined by the size of the sample. Based on this distribution, we can state in probability terms our confidence that our sample results lie within a given margin of error of the truth about the population. If the sample is a *simple random sample* (the formal version of drawing names from a hat) and the population is large, the recipes for confidence intervals learned in elementary statistics apply.

Good data are as much a product of human skill as hybrid corn and compact disc players.

In practice, samples of large, geographically dispersed human populations use sampling designs more complex than a simple random sample. It is common, for example, to divide the population into relatively homogeneous groups (called *strata*) and to select at random separately within each stratum rather than rely on the all-at-once random selection of a simple random sample. More complex designs require correspondingly more complex analysis of the data. Nonetheless, a systematic design and random selection preserve the essential advantages of planned data production over haphazard methods: unbiased estimation with precision that can be known and controlled by the design. The art of sampling human populations does not end with good statistical design, as Chapter 4 emphasizes; but good statistics is the foundation of the art.

Experimentation

Sample surveys are a species of observational study; data are collected without any attempt to interfere with the system observed. *Experiments*, on the other hand, impose some treatment on experimental subjects in order to observe the results. Experiments can study the effects of specific factors of interest, alone and in combination. Well-designed experiments are the only way to obtain fully persuasive evidence of cause-and-effect relations. For these reasons, experimentation is the preferred mode of producing data about causal relations. When direct experimentation is not possible, as in studying the link between smoking and lung cancer in humans, conclusions about causation require large amounts of indirect evidence and remain more subject to challenge than experimental conclusions. On the other hand, experiments often lack the direct link to a population of interest that is present in sample surveys. Experimental subjects may not be a random sample from the population of interest (as when psychologists use college students as subjects but draw conclusions about all adults), and experimental treatments may not be realistic (as when process engineers design a reaction in a small pilot plant). Critics may then dispute the generality of the experimental results.

The design of an experiment, in the statistical sense, describes what treatments are imposed on the subjects and how the available subjects are allocated among these treatments. In the isolated environment of a laboratory, experiments often have a simple design such as

$$\text{Treatment} \longrightarrow \text{Observation} \qquad (1)$$

The design is simple even though the treatment may be very complex. In the field, or when human subjects are involved, such simple designs often produce invalid data because external influences cannot be excluded. The apparently favorable response to a new medical therapy may be merely the placebo effect, the effect of even a dummy treatment administered with authority. The excellent yield of a new soybean variety may be due largely to good weather and the careful agronomic practices of the research farm.

The remedy for the confusion (statisticians call it *confounding*) of treatment effects with external influences is comparison—some patients are given the new drug, while others receive a placebo or a standard drug; some

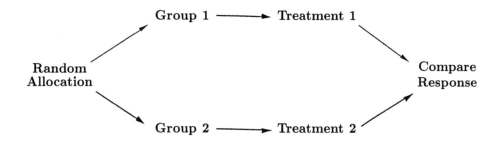

Figure 3: The basic randomized comparative experimental design for comparing two treatments.

plots of land are planted with the new soybean variety, while adjoining plots are planted with a standard variety. If we can obtain equivalent groups and treat them identically except for the experimental treatments, external influences act on all groups and (on the average) have no selective effect on the outcome. Comparative experimentation has been used in agronomy research since the eighteenth century. By the early twentieth century, it was realized that data from many small plots were preferable to comparing the average yield from two large plots. Handbooks presented ingenious arrangements of small plots to protect against systematic fertility differences in any direction across the test field. (Cochran 1976 gives an account of the history.) It remained for R. A. Fisher to realize in the 1920s that random assignment of plots or patients to the treatments would create groups that were (on the average) equivalent in all respects, even those not recognized by the experimenters. Figure 3 displays the simplest version of Fisher's creation, a *randomized comparative experimental design* for comparing two soybean varieties or two medical therapies:

Randomization and comparison in an experimental design, like random selection in a sampling design, avoid bias and allow a probability-based analysis of the resulting data. The analysis compares the observed difference in outcomes to the distribution of differences that would arise in repeated random allocations; differences larger than could reasonably arise by chance are called "statistically significant." The sensitivity of the experiment to real differences among the treatment effects is increased by enlarging the groups, so that the chance variation among them is smaller.

The design in Figure 3 is called *completely randomized*,

because all of the patients or plots are allocated among all of the treatments in a single randomization. Completely randomized designs are the analog of simple random samples; indeed, each treatment is assigned to a simple random sample of the experimental subjects. Just as in sample surveys, more complex designs are common in practice. If the subjects differ in some way that affects response to the experimental treatments, we may first divide the subjects into relatively homogeneous groups and then assign subjects to treatments at random separately within each group. These groups are called *blocks* in experimentation, though they serve the same purposes as strata in survey sampling. The designs used in practice are also complicated by the fact, stressed in Chapter 5, that it is generally more efficient for an experiment to investigate the effects of more than one factor.

Designed data collection imposes an order on the data that can be described by a mathematical model. Once we have a mathematical model, we have reached the threshold of formal statistical inference.

4. INFERENCE

Statistical inference, drawing conclusions from data to some wider universe, builds on data analysis and data production. The numerical measures used to describe data also serve for inference. Statistical designs for producing data are intended in part to ensure that routine inference procedures are justified. Inference applies standard methods of reasoning, relatively few in number (at least in routine settings), to a variety of mathematical models for the phenomena that generate the data. The models include, and the methods of reasoning employ, probability as a description of uncertainty. Conceptual understanding of statistical inference therefore rests on conceptual understanding of probability.

The dependence of formal inference on probability is unfortunate from the teacher's point of view, because probability is perhaps the most difficult to understand of all the topics of elementary mathematics. Garfield and Ahlgren (1988) survey a large body of education research that makes just this point and attempts to explore why a conceptual understanding of probability principles remains so weak, even among students who can do formal exercises in the mathematics of probability. They conclude that, "In spite of this research, however, teaching a conceptual grasp of probability still appears to be a very difficult task, fraught with ambiguity and illusion." The evidence for students' failure to comprehend probability concepts and inferential reasoning based on probability is so strong—and most teachers would add their anecdotal experience to this evidence—that Garfield and Ahlgren suggest exploration of "how useful ideas of statistical inference can be taught independently of technically correct probability."

Certainly the present trend toward reemphasizing actual experience with data analysis in beginning instruction before plunging into probability and inference makes sense pedagogically as well as presenting a more balanced introduction to statistical practice. It is certainly also true that the considerable dose of formal probability traditionally found in beginning statistics courses can be greatly reduced. In this setting, we should limit ourselves to those aspects of probability that are essential for inference. We cannot afford to give probability its due as a subject in its own right. It may be true that more experience with physical chance phenomena and computer simulation will build student understanding of probability. Experience in looking at distributions of data can lead to guided experience with randomness that emphasizes that the tools of data analysis also describe the distributions of outcomes in repetitions of chance phenomena. Perhaps this data-centered approach will allow principles of inference to be taught to users without much in the way of formal probability. (The book by Box, Hunter, and Hunter 1978, which is widely used in engineering and industrial instruction, takes a similar approach.) Yet teachers of statistics, however we face the pedagogical obstacles posed by the difficulty of probability ideas, are obligated to present at least the basic reasoning of confidence intervals and significance testing as essential parts of our subject.

Classical Inference and Probability

The reasoning of classical statistical inference is based on probability as a description of the long-term behavior of chance phenomena. This "frequentist" interpretation views probability as the mathematical idealization of the fact that in many independent trials the proportion (relative frequency) of trials on which an event occurs seems to approach a fixed limit. We observe that some phenomena are *random* in this sense. Although individual outcomes are uncertain, a stable pattern or distribution of outcomes emerges from many repetitions. Heredity and radioactive decay are natural random phenomena, and the deliberate use of chance mechanisms in data collection introduces randomness into the resulting data. Randomness is not a synonym for haphazard behavior. Quite the opposite, it is a kind of order that is susceptible to mathematical description and provides a basis for scientific work, as in heredity and radioactivity. Statisticians use specific chance mechanisms in designs for data production to introduce order into the data as a basis for inference. It is a strength of the frequentist interpretation that probability is seen first as a description of the actually observed phenomenon of randomness, so that there is a natural tie to empirical science. The accompanying weakness of this point of view is that the range of phenomena that can be described by probability is limited to those that can in principle be indefinitely repeated.

The considerable dose of formal probability traditionally found in beginning statistics courses can be greatly reduced. In this setting, we should limit ourselves to those aspects of probability that are essential for inference. We cannot afford to give probability its due as a subject in its own right.

The use of probability models interpreted in the frequentist sense distinguishes classical statistical inference from both non-model-based data analysis and Bayesian inference based on probability as a measure of degree of belief. It should now be clear why classical inference is most secure when data are produced by a randomized design: Deliberate randomization justifies a probability model for the data and so justifies procedures derived from that model. This justification goes further than is commonly realized. For example, analysis of variance, the classical statistical procedure for comparing the mean responses resulting from several treatments, is usually described as requiring that the data come from normally distributed populations. But in fact, data produced by a randomized experimental design require no such distributional assumption. The randomization itself provides the basis for a test of the hypothesis that there are only random differences among the responses, and the usual analysis-of-variance test is a continuous approximation to this discrete test. See Pitman (1937, 1938) or Edgington (1987).

Inference is, of course, often applied to data that are not the product of randomized designs. In such circumstances, we must be willing to assume that some probability model does nonetheless govern the data production.

> **Example.** Control charts and other procedures for statistical process control often assume that successive measurements are independent observations on a distribution that represents the state of the process. The distribution may be, for example, normal if some characteristic of the product is measured, or Bernoulli if each product is simply categorized as meeting specifications or not. The essential aspect of the model for many purposes is the independence of successive observations. Independence often holds (at least approximately) for a process manufacturing discrete parts. Measurements on a continuous-flow process, on the other hand, often show serial correlation unless well separated in time, so that a more elaborate probability model is required. Lack of independence can change not only the probability properties of control charts but even the qualitative principles that users rely on to understand statistical quality control. MacGregor (1990) gives a simple discussion of some of the qualitative effects of dependence.

The appropriateness of a model that is assumed to hold but not justified by the design of the data production can be checked from the data themselves (diagnostics again). Even so, extensive investigation can leave issues such as independence undecided. Are sequences of hits and misses in shooting a basketball the chance results of independent trials, or does the "hot hand" lead to streaks of shots made? Tversky and Gilovich (1989a,b) and Larkey, Smith, and Kadane (1989) argue the oppos-

ing sides of a continuing debate about this question. In the case of the hot hand, substantive conclusions about the psychology of competition are at issue. In process control, the issue is the appropriateness and performance of statistical methods that assume independence. Both examples remind us of the difficulty of specifying a model and indirectly of the advantages of designed data production.

In yet other situations, it is possible to specify a model (at least in part) based on a substantive scientific understanding of, for example, the kinetics of a chemical reaction or the spread of an epidemic. For a discussion of this and other aspects of the specification of models in statistics, see the side-by-side papers of Cox (1990) and Lehmann (1990). Our point here is more basic: formal inference proceeds from a specified model that has a probabilistic component.

This basic point has led some critics, most prominently W. Edwards Deming, to deny the applicability of formal inference except in "enumerative studies" that actually sample from a physically existing population. (See Deming's brief but unyielding remarks in Deming 1988, page 104.) Although this critique of inference grasps the importance of models, it is far too pessimistic. Models for processes arise in many ways. That those based on actual randomization are most secure does not imply that no others are useful.

Confidence Intervals

Classical inference attaches to its conclusions a probability statement that expresses our confidence in *the method* used to draw the conclusion. That probability statements concern the inference method rather than the particular conclusion reached from one set of data is central to the reasoning of classical inference. The reasoning is clearest for *confidence intervals*. We wish to estimate from data some unknown parameter, call it θ, of the underlying population or process. The data produce an estimate of the unknown θ, call it T. It is not satisfactory to report simply that T is our estimate of θ, for we have given no indication of the uncertainty in our result. A confidence interval procedure has two parts: a recipe for computing an interval from the sample data, and a confidence level, the proportion of all samples for which the interval will cover the true value of θ. The confidence level is a probability in the frequentist sense. It is obtained from the distribution that describes the variation of the estimator T in repeated sampling.

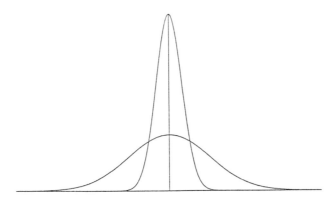

Figure 4: The distribution of the mean of 10 observations compared with the distribution of a single observation.

Consider a simple specific setting. We wish to estimate the mean μ of a population or process. Common sense suggests the sample mean $\bar{x}$ as an estimator of μ. We know that repeated samples from the same population or process would produce different values of $\bar{x}$. Classical inference asks explicitly "How will $\bar{x}$ vary in repeated samples from the same population?" In general settings, we can answer this question to any desired accuracy by simulation, and often to good accuracy by theory. In some simple settings, theory gives an exact answer. Suppose that we are content to model our data as n independent observations from a normal distribution with unknown mean μ and known standard deviation σ. Theory then tells us that $\bar{x}$ will vary according to a normal distribution with the same mean μ as the population but with smaller standard deviation $\sigma/\sqrt{n}$. This is the *sampling distribution* of $\bar{x}$. Figure 4 displays the distribution of a single normal observation and the sampling distribution of $\bar{x}$ from a sample of 10 observations.

Statistical theory now does its work, building on the sampling distribution. First, theory confirms our common-sense choice of $\bar{x}$ to estimate μ by demonstrating that $\bar{x}$ has many desirable properties as an estimator of μ. For example, $\bar{x}$ is unbiased (has expected value equal to the unknown μ) and in fact has the smallest variance among all unbiased estimators of μ. These properties already assume that performance is to be assessed from repeated use. Unbiasedness says that $\bar{x}$ is correct on the average in the long run, though it will over- or underestimate μ in almost all individual instances. What is more, we immediately obtain the usual recipe $\bar{x} \pm 1.96\sigma/\sqrt{n}$ for a 95% confidence interval from the fact that 95% of any normal distribution lies within

1.96 standard deviations of the mean. The phrase "95% confidence" refers to the probability that this random interval will contain the true μ. Applied to a specific outcome, 95% confidence means "I got this result by a method that is correct 95% of the time in repeated use."

The idea of a sampling distribution—the distribution of values taken by a statistic such as $\bar{x}$ in repeated data production—is central to the reasoning of classical inference. Students find the idea difficult, so that careful explanation and demonstration by simulation are needed. In fact, the conceptual difficulties begin even before we perform the thought experiment of repeating our data production many times. The basic normal probability model itself is often shorthand for a train of thought that we fail to make clear to students: Data analysis reveals that the distribution of many observations from this population is at least roughly described by a normal curve; consider choosing one observation at random; the probability distribution of a single observation is the same normal distribution that describes the population of all observations. The transition from a normal distribution as a description of the overall shape of a distribution of data to a probability distribution is made via the thought experiment of choosing a single item at random.

We ought to ask, with Garfield and Ahlgren, whether formal probability theory is essential or even helpful in overcoming the conceptual difficulties associated with "population distribution" and "sampling distribution." A more empirical approach via simulation experiments and data analysis of the actually observed distributions is certainly promising.

Bayesians deny the relevance of the sampling distribution, noting that it is a distribution of values that were not in fact observed. They also note that making a probability statement about the method rather than about the actual result is a bit indirect and not easily understood by users. Yet many users, such as polling organizations that take weekly random samples of the adult population, find it quite satisfactory to express the uncertainty of their estimates by giving a margin of error that will be attained in 95% of these samples in the long run.

Tests of Significance

The other common method of classical inference, *tests of significance* or *hypothesis testing*, is more controversial. Statisticians disagree over the proper interpretation of tests. Most statisticians do agree that tests are

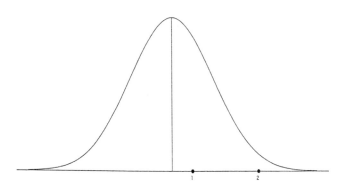

Figure 5: If observations vary according to this normal curve, observation 2 casts doubt on the hypothesis that the mean is as marked, whereas observation 1 does not.

overused. How *large* an effect is (information not given by a test alone) is usually more important than how statistically significant it is. Yet the basic purpose of statistical tests is straightforward and clearly necessary. We see in our data some striking effect. Patients treated with the new drug are living longer, or the rate of customer complaints has dropped following redesign of the product. Of course, we expect patient lifetimes and customer complaint rates to vary. Are the changes we see real in the sense of being too large to be plausibly due to chance variation? *Statistically significant* means just that—the observed effect is too large to occur often by chance alone—and no more. In particular, since significance does not report the size of an effect, significant effects can be too small to be of any practical importance.

A formal procedure to assess whether an observed effect is too large to be easily attributed to chance is essential because our informal judgment is very bad. Many studies (e.g., Tversky and Kahneman 1971) have shown that we all too easily attribute chance patterns in data to some substantive cause. Runs in gambling, clusters of neighboring cancer cases, and many other patterns are much more likely to occur by chance than our intuition can concede. Significance tests are like an engineer's calculation of the load on the beam above our heads: The beam looks strong enough, but we feel more secure knowing that calculation verifies appearance. The fallibility of our judgment of pattern and variation is also a reminder of the limitations of data analysis not accompanied by inference.

The essential method of tests of significance is also

straightforward: Compare the result actually observed to a reference distribution that displays the chance variation expected in our setting. If our result is extreme in the direction suggested by a substantive hypothesis about the underlying process, we have good evidence of a real effect. If our result lies within the range that would often occur by chance, we lack good evidence of a systematic effect. In practice, the reference distribution is the sampling distribution of an appropriate statistic. In many common situations, we can center our measurements so that zero represents "no change" or "no effect." We may, for example, assess the yield of a new process and a standard process in several runs each, then measure the improvement by the difference of the observed means, new minus standard. Figure 5 describes this situation. The normal curve represents the sampling distribution of the observed mean difference, assuming that the two processes in fact have the same mean yield. Values sufficiently far above zero, like point 2 in Figure 5, give evidence that the expected chance variation has been disturbed by some systematic effect. If our data came from a randomized comparative experiment (but not necessarily otherwise), we can confidently conclude that the systematic effect must be the higher mean yield of the new process.

It is a short step from this informal reasoning to more formal statements based on probability. The one-sided probability, calculated from the reference distribution, that chance would produce an observation at least as large as point 1 in Figure 5 is 0.31. An observation this large would often occur by chance. The corresponding tail probability for the second observation is 0.017. Because chance variation would produce an observation this large fewer than one time in fifty, we have evidence that the new process does have a higher mean yield.

The translation of this common-sense reasoning into formal tests by stating null and alternative hypotheses, specifying a test statistic, and comparing an observed value of this statistic to an appropriate distribution is notoriously hard to follow. To be fair, the common-sense description conceals many difficulties. It is couched in terms of degree of evidence for a systematic effect, expressed as the probability that an effect at least as strong as that observed would occur by chance. This probability is the *P-value*. Statistical software reports a *P*-value when asked to perform a test, and (perhaps in part for this reason) the "strength of evidence" reasoning seems to be favored by most working statisticians. An alternative, backed by the Neyman-Pearson theory taught in statistical theory courses, regards testing as a "two-action decision problem" that asks us to decide between the null and alternative hypotheses. The Neyman-Pearson approach presents us with two competing hypotheses and emphasizes the two types of error that decisions can produce. We make a *Type I error* in rejecting a true null hypothesis and a *Type II error* in failing to reject a false null hypothesis. The two error probabilities, like confidence levels, describe the performance of the inference method.

The decision approach leads to the attempt to find tests that are optimal by specific criteria. The most common criterion seeks tests that minimize the probability of Type II error (maximize power) for a fixed probability of Type I error (significance level). These accomplishments remind us that an emphasis on P-values does not naturally accommodate considerations of power, which are clearly of practical importance, or the search for optimal procedures. The Neyman-Pearson emphasis on decisions, on the other hand, does not fit most applied inference problems. What is more, the emphasis on Type I error needed to obtain optimal tests does not arise within the decision orientation, but reflects the influence of the strength-of-evidence approach. Thus neither approach is fully satisfactory, and aspects of the two are usually mixed in practice. Indeed, the brief definitions of Type I and Type II errors given above reflect that mixture in their emphasis on the null hypothesis. When two distinct modes of reasoning are mixed with varying emphases, it is not surprising that disagreement and even confusion are common. Undergraduate instruction must usually oversimplify to convey some basic concepts. Instruction driven by practice will favor the P-value approach.

Bayesian Inference

The principal alternative to classical ideas in the field of probability-based inference is offered by Bayesian approaches. Bayesians extend the probability models that are the basis of inference to include prior information about the values of unknown parameters. A product engineer, for example, does not know the mean time to failure of a new mechanical system. But from experience with similar products, she is convinced that the mean time lies between 50 and 200 hours and that values near 150 hours are most probable. The engineer's prior knowledge and informed judgment can be expressed as a probability distribution for the unknown mean failure time. Bayesian inference methods combine such prior information with data (the observed times to failure of

units placed on test) to estimate the mean failure time for all such systems.

Formally, the Bayesian approach regards the unknown parameter θ as the value of a random variable Θ with known *prior distribution* $p(\theta)$. The data are generated by a probability model $f(x|\theta)$ that is now understood to describe the conditional distribution of the observations X given that $\Theta = \theta$. From this enlarged model, we obtain the conditional distribution of Θ given the observed data x, $p(\theta|x)$. Inference is based on this *posterior distribution*. The computation of the posterior distribution in elementary settings uses Bayes's theorem from basic probability theory, hence the name "Bayesian."

Most statisticians agree that Bayesian methods should be employed when it makes sense to regard an unknown parameter θ as the value of a random variable with known distribution and when this information is relevant to the problem. Bayesians assert that these conditions always hold. In order to maintain this position, it is necessary to extend the frequentist concept of probability. Probability is often used informally to express degrees of personal belief: "I think the Giants have a 20% chance to win the Superbowl this year." Such *subjective probabilities*, if they are to be consistent, must obey the axioms of the usual mathematical theory of probability. The mathematics is not affected by our interpretation. Subjective or personal probabilities can be based on frequency when events can be indefinitely repeated, but they also apply to unrepeatable events such as winning the Superbowl this year. The prior distribution required by the Bayesian model is often a subjective distribution that expresses partial knowledge and informed judgment. Bayesian inference, being based on a broader interpretation of probability, has a wider scope than classical inference. Moreover, the posterior distribution allows us to make probability statements about the specific conclusion at hand, not just about the long-run performance of our methods. These statements are easier for users to interpret than confidence intervals and significance tests. Bayesian reasoning offers some substantial advantages. Teachers and users of statistics need a basic acquaintance with Bayesian methods. Lee (1989) and Press (1989) are elementary expositions of the Bayesian approach.

The Bayesian position remains controversial because most working statisticians remain unconvinced that the additional information required by these methods is always available and relevant. It is not easy in practice to effectively extract a specific prior distribution from a decision-maker whose judgment is conceded to be informative. Different individuals may have quite different prior opinions, and in a frequentist setting all of these individual opinions may differ greatly from the frequencies that will actually be observed in repeated trials. Recall that our conceptual grasp of random behavior is very deficient. Although diagnostic methods that allow the data to criticize the usual statistical models are well-advanced, it is inherently much harder to criticize from the data a prior distribution for the parameter. The Bayesian model, which gains its greater scope from willingness to include subjective information, remains more subjective than classical models in the specific sense of being less easily criticized and falsified from data. (Of course, large amounts of data will in all common settings eventually overwhelm the effect of the prior distribution on Bayesian inference, a *de facto* criticism of the model by the data.) Even when subjective prior probabilities are available, many persons deny their relevance in estimating, for example, gene frequencies or the measurement bias of a new instrument.

There are, I think, good reasons not to stress Bayesian methods in beginning instruction about inference.

The arguments just outlined for and against Bayesian ideas are oversimplified and would be disputed by both sides. Some Bayesians have responded to criticism by modified approaches such as seeking Bayesian procedures that are not sensitive to changes in the prior distribution and so offer the advantages of Bayesian conclusions without the necessity of assuming that we know the correct prior distribution. Berger (1984, 1990) advocates this *robust Bayesian* approach. Barnett (1982) offers an overview and comparison of the competing philosophies of inference. Box (1983) makes an effective plea for eclecticism or ecumenism. The problem of inductive inference from uncertain data is, after all, a knotty philosophical issue. It is not surprising that statisticians do not have a single agreed-upon approach, nor that most practicing statisticians are eclectic in their choice of methods that are effective in specific settings.

There are, I think, good reasons not to stress Bayesian methods in beginning instruction about inference. First, they require a firm grasp of conditional probability. As Garfield and Ahlgren (1988) note, conditional probability is a particularly difficult idea. Even able students often fail to distinguish $P(A|B)$, $P(B|A)$, and $P(A \text{ and } B)$

in verbal discussion. To follow Bayesian reasoning, a student must understand the distinction between the conditional distribution of the statistic given the parameter and the conditional distribution of the parameter given the actually observed value of the statistic. This is fatally subtle. In addition, although the subjective interpretation of probability is quite natural, it diverts attention from randomness and chance as observed phenomena in the world whose patterns can be described mathematically. An understanding of the behavior of random phenomena is an important goal of teaching about data and chance; probability understood as personal degree of belief is at best irrelevant to achieving this goal. The line from data analysis through randomized designs for data production to inference is clearer when classical inference is the goal.

5. THE ROLE OF STATISTICAL THEORY

The outline of basic statistical ideas just presented was driven by the nature of statistical practice rather than by the mathematical theory of statistical inference. The stress placed on data analysis and on designs for producing data is an aspect of this practical emphasis. So is the concentration on the role of models in inference, on diagnostics as an essential part of statistical work, and on the underlying modes of reasoning in inference. This emphasis should not obscure the fact that mathematical analysis has been of inestimable value in statistics, as in other quantitative sciences. Here are a few of the accomplishments of statistical theory in areas touched on in our exposition.

- For a given model, we can derive the sampling distribution of various statistics and the properties of inference procedures. We can often find procedures that are optimal by some appropriate criterion. An excellent example is the theory of the *general linear model* that underlies such classical procedures as regression and analysis of variance. An elegant structure that combines projections in finite-dimensional Euclidean space and multivariate normal probability distributions allows us to derive and justify statistical procedures and obtain the probability results needed to apply them in a general setting. The sampling distribution and minimum-variance-unbiased property of the sample mean $\bar{x}$ referred to in an earlier example are very special cases of this general theory.

- Data analysts observe that, for example, the median is much more resistant to outlying observations than is the mean. Mathematical study can quantify resistance and explicitly compare the resistance of more complicated statistics. Even though the tools of data analysis are often not based on a mathematical model for the data, mathematical study still clarifies their advantages and helps us choose among competing methods.

- Mathematical analysis demonstrates that in most settings the effect of the prior distribution on Bayesian inference vanishes as the number of observations increases, and that Bayesian and classical methods converge to similar conclusions in the limit. It is satisfying that enough data drive competing modes of reasoning toward common conclusions.

The Limitations of Theory

Nonetheless, introductory instruction that presents statistics as if it were mathematics gives an inadequate picture of the field. More important, elementary instruction that is almost entirely non-mathematical but is guided by the theory lurking in the background is similarly misleading. To see why, consider a simple setting discussed in all beginning statistics courses, *two-sample problems*. This setting is important in practice, and most of the comments I make about it apply to the much wider class of methods based on the general linear model. The mathematical model for two-sample problems states that the data are independent random samples from two normal populations:

$$X_1, X_2, \ldots, X_n \quad \text{iid} \quad N(\mu_1, \sigma_1)$$

$$Y_1, Y_2, \ldots, Y_m \quad \text{iid} \quad N(\mu_2, \sigma_2)$$

Statistical theory allows us to obtain tests and confidence intervals in this setting, and guarantees that they have some desirable properties. This is valuable. Why isn't it the whole story?

First, *the mathematical model is incomplete*. It does not distinguish between observational and experimental studies, a distinction fundamental to science and to the validity of many conclusions drawn from statistical analyses. The model, and inference methods based on it, are the same for observational and experimental data. The superiority of experiments in justifying conclusions about cause-and-effect is not visible in the mathematics.

Second, *the theoretical merit of a statistical procedure is not the same as its practical merit*. In the two-sample setting, we may wish to compare the centers of the two

populations as described by the means μ_1 and μ_2. Or we may wish to compare their variability, described by the standard deviations σ_1 and σ_2. Evidence of unequal means would be assessed by the two-sample t test, evidence of unequal σ's by the F ratio test. In theory, these tests are of equal merit. Both are likelihood-ratio tests. That is, they are products of an honored organizing principle of statistical theory and share certain large-sample optimality properties. Beginning instruction that is driven by theory often presents the t and F tests on the same level. Because the theoretically cleanest version of the two-sample t requires the assumption that $\sigma_1 = \sigma_2$, some texts even suggest that an F test be conducted as a preliminary to the t test. More generally, tests of equality of variances are often recommended as preliminaries to a comparison of means by analysis of variance.

Introductory instruction that presents statistics as if it were mathematics gives an inadequate picture of the field. More important, elementary instruction that is almost entirely non-mathematical but is guided by the theory lurking in the background is similarly misleading.

Similar theory, however, does not imply similar value in practice. The two-sample t for comparing means is a most useful tool, but the F ratio for comparing variances is almost worthless. (Note that the F tests encountered in analysis of variance compare two or more *means*; they are analogues of the two-sample t test and share its usefulness. It is formal tests for equality of two or more *variances* that join the F ratio in lack of practical value.) The reason for this important distinction is that no data are exactly normal. A procedure useful in practice must therefore be somewhat *robust*, that is, not sensitive to mild departures from normality. The two-sample t is quite robust; the F ratio is so sensitive to even small departures from normality as to be almost useless.

Pearson and Please (1975) is one of many sources that report abundant evidence contrasting the robustness of t and F. These authors carried out a large simulation study of the actual significance level attained by t and F tests conducted at the nominal 5% level on data with various nonnormal distributions. The true significance level of the two-sample t test in samples of size 25, for example, remained between 4% and 6% for the entire range of distributions simulated. For these same distributions, the F ratio had significance levels ranging from 1% to 11%. Such a test is hardly useful. As a check on the equal-variances assumption of the analysis-of-variance test for equality of means, it is silly. As Box (1953) remarks, "To make a preliminary test on variances is rather like putting to sea in a rowing boat to find out whether conditions are sufficiently calm for an ocean liner to leave port."

My primary point is that theory is an imperfect guide to practice. But in light of the failure of the F test, what ought we to teach our students about inference on spreads? In beginning instruction, nothing. There are no simple formal inference procedures that can be recommended in practice. (Nonparametric tests for spread have fatal problems also.) At an advanced level, more can be said. Modern jackknife and bootstrap methods (Boos and Brownie 1989) work well in moderately large samples. Like many newer procedures, these are very computer-intensive. For smaller samples, Conover et al. (1981) discovered via a large simulation study some practically useful methods.

Finally, *theoretically-based procedures may require unrealistic assumptions.* Not all unrealistic assumptions are barriers to practical use, as the robustness against non-normality of the t procedures reminds us. But the version of the two-sample t test most often found in texts on statistical theory requires another assumption: Not only do the two populations being compared have normal distributions, but their standard deviations σ_1 and σ_2 are equal. The test statistic derived in theoretical treatments of this setting is the pooled-sample t statistic

$$T_p = \frac{\overline{X} - \overline{Y}}{s_p\sqrt{\frac{1}{n} + \frac{1}{m}}}$$

Here s_p is an estimate of the common standard deviation obtained by pooling estimates from the two samples. The pooled-sample t test is a likelihood ratio test, and the statistic T_p has a t distribution under the null hypothesis of equal population means. This test is somewhat robust against unequal σ's if the sample sizes are equal, but not otherwise. One study of the actual significance level of T_p using 5% t critical points found for $n = 5$ and $m = 15$ that the true significance level ranged from 0.24% to 31.7% for normal populations as σ_1/σ_2 varied. Moreover, we have just seen that the assumption $\sigma_1 = \sigma_2$ is difficult to verify.

The natural two-sample t statistic is

$$T = \frac{\overline{X} - \overline{Y}}{\sqrt{\frac{1}{n}s_1^2 + \frac{1}{m}s_2^2}}$$

The statistic T does *not* have a t distribution under the null hypothesis and is not supported by any theoretical optimality properties. But its critical points can be very closely approximated by those of a t distribution with degrees-of-freedom, usually not an integer, determined from the data by a somewhat messy recipe. A substantial literature (e.g., Leaverton and Birch 1969, Scheffé 1970, and Best and Rayner 1987) demonstrates the accuracy of this approximation for even quite small samples, and demonstrates in addition that when in fact $\sigma_1 = \sigma_2$, using T sacrifices very little power relative to T_p.

Statistical software has no difficulty computing the degrees of freedom for T and tail probabilities for t distributions with non-integer degrees of freedom. Procedures based on the statistic T are available in all standard software packages, and often are the default two-sample t procedures. The rather special pooled-sample procedures are less often used in practice. They ought not to prevail in teaching simply because they are rooted in theory.

6. CONCLUSION

Statistics is a mathematical science, but it is not a branch of mathematics. Statistics is a methodological discipline, but it is not a collection of methods appended to economics or psychology or quality engineering. The historical roots of statistics lie in many of the disciplines that deal with data; its development owes much to mathematical tools, especially probability theory. But by the mid-twentieth century statistics had clearly emerged as a discipline in its own right, with characteristic modes of thinking that are more fundamental than either specific methods or mathematical theory. We can summarize the core elements of statistical thinking as follows:

1. The omnipresence of **variation** in processes. Individuals are variable; repeated measurements on the same individual are variable. The domain of a strict determinism in nature and in human affairs is circumscribed.

2. The need for **data** about processes. Statistics is steadfastly empirical rather than speculative. Looking at the data has first priority.

3. The design of **data production** with variation in mind. Aware of sources of uncontrolled variation, we avoid self-selected samples and insist on comparison in experimental studies. We introduce planned variation into data production by use of randomization.

4. The **quantification** of variation. Random variation is described mathematically by *probability*.

5. The **explanation** of variation. Statistical analysis seeks the systematic effects behind the variability of individuals and of measurements.

Statistics is a mathematical science, but it is not a branch of mathematics. Statistics is a methodological discipline, but it is not a collection of methods appended to economics or psychology or quality engineering.

The higher goal of teaching statistics is to build the ability of students to deal intelligently with variation and data. Whatever our audience, whether we are focusing on theory or on methods, we ought not to lose sight of that goal.

Acknowledgments

Writing of this article received partial support from the National Science Foundation through grant DMS-8901922 to Purdue University. The author is grateful to Lynne Hare, David Hoaglin, Lincoln Moses, Glenn Shafer, and Judith Tanur for helpful comments on an earlier draft.

David S. Moore is Professor of Statistics at Purdue University. He received an A.B. from Princeton in 1962 and a Ph.D. in mathematics from Cornell in 1967. His professional interests include tests of fit, analysis of categorical data, and engineering statistics in addition to a long-standing commitment to statistical education. He coordinates instruction in Purdue's statistics department and has won a university-wide teaching award. Professor Moore has served on the editorial boards of *Technometrics* and the *Journal of the American Statistical Association* and as statistics program director at the National Science Foundation. He is a fellow of both the American Statistical Association and the Institute of Mathematical Statistics and an elected member of the International Statistical Institute.

REFERENCES

Barnett, V. (1982), *Comparative Statistical Inference*, second edition, New York: John Wiley.

Berger, J. O. (1984), "The Robust Bayesian Viewpoint" (with discussion), in *Robustness of Bayesian Analyses*, ed. J. B. Kadane, Amsterdam: North-Holland, 63–144.

—— (1990), "Robust Bayesian Analysis: Sensitivity to the Prior," *Journal of Statistical Planning and Inference*, **25**, 303–328.

Best, D. J. and Rayner, J. C. W. (1987), "Welch's Approximate Solution for the Behrens-Fisher Problem," *Technometrics*, **29**, 205–210.

Boos, D. D. and Brownie, C. (1989), "Bootstrap Methods for Testing Homogeneity of Variances," *Technometrics*, **31**, 69–82.

Box, G. E. P. (1953), "Non-Normality and Tests on Variances," *Biometrika*, **40**, 318–335.

—— (1983), "An Apology for Ecumenism in Statistics," in *Scientific Inference, Data Analysis, and Robustness*, eds. G. E. P. Box, T. Leonard, and C.-F. Wu, New York: Academic Press, 51–84.

Box, G. E. P., Hunter, W. G., and Hunter, J. S. (1978), *Statistics for Experimenters*, New York: John Wiley.

Cochran, W. G. (1976), "Early Development of Techniques in Comparative Experimentation, " in *On the History of Statistics and Probability*, ed. D. B. Owen, New York: Marcel Dekker, 3–25. The paper also appears in Cochran's *Contributions to Statistics*, (1982), New York: John Wiley.

Conover, W. J., Johnson, M. E., and Johnson, M. M. (1981), "A Comparative Study of Tests for Homogeneity of Variances, with Applications to the Outer

Continental Shelf Bidding Data," *Technometrics*, **23**, 351–361.

Cox, D. R. (1990), "Role of Models in Statistical Analysis," *Statistical Science*, **5**, 169–174.

Deming, W. E. (1988), "Comment: Recollections about Harold Hotelling," *Statistical Science*, **3**, 103–104.

Edgington, E. S. (1987), *Randomization Tests*, second edition, New York: Marcel Dekker.

Efron, B. (1979), "Computers and the Theory of Statistics: Thinking the Unthinkable," *SIAM Review*, **21**, 460–480.

Garfield, J. and Ahlgren, A. (1988), "Difficulties in Learning Basic Concepts in Probability and Statistics: Implications for Research," *Journal for Research in Mathematics Education*, **19**, 44–63.

Jones, L. V. (ed.) (1986), *The Collected Works of John W. Tukey: (Vol. III) Philosophy and Principles of Data Analysis, 1949–1964; (Vol. IV) Philosophy and Principles of Data Analysis, 1965–1986*, Monterey, CA: Wadsworth & Brooks/Cole. A reviewer recommends paper 12 in Volume IV as a starting point.

Larkey, P. D., Smith, R. A., and Kadane, J. B. (1989), "It's OK to Believe in the 'Hot Hand,'" *Chance*, **2**, number 4, 22–30.

Leaverton, P. and Birch, J. J. (1969), "Small Sample Power Curves for the Two Sample Location Problem," *Technometrics*, **11**, 299–307.

Lee, P. M. (1989), *Bayesian Statistics: An Introduction*, New York: Oxford University Press.

Lehmann, E. L. (1990), "Model Specification: The Views of Fisher and Neyman, and Later Developments," *Statistical Science*, **5**, 160–168.

MacGregor, J. F. (1990), "A Different View of the Funnel Experiment," *Journal of Quality Technology*, **22**, 255–259.

Mickey, M. R., Dunn, O. J., and Clark, V. (1967), "Note on the Use of Stepwise Regression in Detecting Outliers," *Computers and Biomedical Research*, **1**, 105–111.

Minitab Inc. (1989), *Minitab Reference Manual, Release 7*, State College, PA: Minitab Inc.

Nielsen, A. C. Jr. (1986), "Statistics in Marketing," in *Making Statistics More Effective in Schools of Business*, eds. G. Easton, H. V. Roberts, and G. C. Tiao, Chicago: University of Chicago Graduate School of Business.

Nisbett, R. E., Fong, G. T., Lehman, D. R., and Cheng, P. W. (1987), "Teaching Reasoning," *Science*, **238**, 625–631.

Pearson, E. S. and Please, N. W. (1975), "Relation between the Shape of Population Distribution and the Robustness of Four Simple Test Statistics," *Biometrika*, **62**, 223–241.

Pitman, E. J. G. (1937), "Significance Tests Which May Be Applied to Samples from Any Populations," *Journal of the Royal Statistical Society, Supplement*, **4**, 119–130.

Pitman, E. J. G. (1938), "Significance Tests Which May Be Applied to Samples from Any Populations. III. The analysis of variance test," *Biometrika*, **29**, 322–335.

Press, S. J. (1989), *Bayesian Statistics: Principles, Models, and Applications*, New York: John Wiley.

Scheffé, H. (1970), "Practical Solutions of the Behrens-Fisher Problem," *Journal of the American Statistical Association*, **65**, 1501–1508.

Tversky, A. and Gilovich, T. (1989a), "The Cold Facts about the 'Hot Hand' in Basketball," *Chance*, **2**, number 1, 16–21.

—— (1989b), "The 'Hot Hand': Statistical Reality or Cognitive Illusion?" *Chance*, **2**, number 4, 31–34.

Tversky, A. and Kahneman, D. (1971), "Belief in the Law of Small Numbers," *Psychological Bulletin*, **76**, 105–110.

Vallone, R. and Tversky, A. (1985), "The Hot Hand in Basketball: On the Misperception of Random Sequences," *Cognitive Psychology*, **17**, 295–314.

Chapter 2

Data Analysis

Paul F. Velleman
Cornell University

David C. Hoaglin
Harvard University

1. WHAT IS DATA ANALYSIS?

As the link between statistics and diverse fields of application, data analysis confronts the challenge of turning data into useful knowledge. Data analysis combines an attitude and a process, supported by well-chosen techniques. The attitude distills the scientist's curiosity about regularity, pattern, and exception. The process iteratively peels off patterns so that we can look beneath them for more subtle (and often more interesting) patterns. The techniques make few assumptions about the data and deliberately accommodate the unexpected.

Data analysis starts with data. The data may come from a known structure, such as a designed experiment or a sample survey; or they may be serendipitous, collected with no particular analysis in mind or with some entirely different analysis in mind. Regardless of their source, we begin by examining the data, usually in graphical displays. At this stage we aim to get a general idea of any patterns while remaining open to unexpected features.

When we recognize a pattern in the data, we attempt to summarize it in a suitable function. This step brings several benefits. First, the function provides a simpler, though incomplete, description. Second, its form often leads to better understanding. Third, by removing a clear pattern (prices have inflated over time, taller people weigh more, higher doses of the drug have a greater effect) we may expose subtler patterns.

Even when a pattern commands attention, data analysis always looks at the *residuals*, the difference at each data point between the observed value and the function describing the pattern. Residuals reveal ways to improve the description of the pattern. They may reveal observations that fail to fit the pattern, which should then re-

ceive special attention (for they often tell us much about the data). Or they may reveal some additional, more subtle, pattern that was hiding beneath the first one and can now be described.

The iterative process of describing patterns, subtracting them, and searching anew for pattern in the residuals continues until the data analyst decides to stop. Although general rules guide this decision, ultimately it is a subjective one.

As the link between statistics and diverse fields of application, data analysis confronts the challenge of turning data into useful knowledge.

Once we have found a satisfactory pattern, we may wish to assess how well it describes the data and how well it might describe additional data values. One philosophy of data analysis, articulated by John W. Tukey (1962, 1977), refers to these two broad phases as *exploratory data analysis* (EDA) and *confirmatory data analysis*. Because other chapters in this volume (Chapter 1, Chapter 9, and especially Chapter 7) discuss aspects of confirmation, we focus primarily on exploration.

To facilitate the processes of exploratory data analysis, Tukey introduced a wealth of innovative techniques and defined many new terms. The immediate usefulness of these techniques was sufficient reason for many to recognize EDA as a major advance in dealing with data. Statistics packages have adopted some of the methods and used some of the terminology, furthering the impression that by using these methods (and intoning the terms as a mantra) one could automatically perform an exploratory analysis.

However, the essence of EDA is its philosophy, an approach to data that differs in fundamental ways from conventional confirmatory statistics (a term that covers the broad range of inferential statistics). And the essence of the EDA philosophy is embodied in how we manipulate and think about data rather than in any particular plots or analyses. Indeed, EDA is most effective when used along with conventional approaches. A careful data analyst uses *both* exploratory *and* confirmatory methods.

2. THE DEVELOPMENT OF A DATA ANALYSIS PHILOSOPHY

Statistics is descended from two disciplines that have interbred over the past two and a half centuries. One is the mathematical study of probability and chance events. The other is the scientific attempt to draw conclusions from data in the face of inevitable error and imprecision. Modern statistics emerged around the beginning of this century in a form that resembled its scientific father more than its mathematical mother.

As statistics matured during the middle of this century, statisticians developed mathematical foundations for most modern techniques. Along the way they developed methods that were optimal in some sense. For example, *maximum-likelihood* methods, which (under customary assumptions) include the most commonly used methods, have many attractive mathematical properties. Unfortunately, real-world data do not satisfy the customary assumptions. Worse, the failure of these conventional assumptions can cause many maximum-likelihood statistics to produce poor estimates, suggest misleading conclusions, and otherwise behave dangerously.

Concern with optimality properties follows a tradition in mathematics that values elegant results for their internal consistency and completeness, but not for their relevance to the real world. By contrast, a scientist would reject even the most elegant theory for failing to describe the observed world. Data analysis resembles science more than mathematics.

Data analysis does not aim simply to answer questions. This orientation is a fundamental departure from formal hypotheses that can be tested to yield a yes or no answer. Instead, data analysis focuses on developing better questions as part of a process. As we learn more about

our data, we are able to formulate better questions and build better models. If we do not insist on seemingly optimal answers, we can continue to make progress in refining our models and learning still more.

3. SOME CONTRIBUTIONS OF DESIGN

Although data analysis methods may differ from traditional methods, they share a central concern for identifying, understanding, and controlling sources of variation. This concern arises in many ways.

Statistical models almost always work best for homogeneous data. The form of the homogeneity usually depends on the model; some models expect observations to be independent and identically distributed, whereas others settle for equal variance across groups or over time, or for other specific consistencies.

Data analysis tools are especially effective at identifying and isolating inhomogeneities in data. These may take the form of anomalous observations, nonconstant variation, or even the recognition that the data divide into subgroups that are internally homogeneous but differ from each other.

Outliers are an abiding concern of data analysis. Observations that deviate from the overall pattern of the data can have a profound effect on fitted models and hence on the interpretation of the data. More important, such extraordinary observations often hold the key to understanding important aspects of the data. As computers have become the principal engine for calculation, the danger of failing to see outliers has grown; a single extraordinary observation can dominate a statistical analysis without being evident in the standard printed results.

Data analysis resembles science more than mathematics.

In experiments and surveys, good design tries to minimize the effects of inhomogeneity. Careful exploration of data from designed experiments can reveal failures of the design (and often helps to make future experiments better). Analysis of serendipitous data, which probably have not come from a designed experiment or survey, can help to "correct the design" by identifying and isolating homogeneous subgroups. These groups can then be explored for pattern with much greater success.

Investigation of outliers and subgroups in data often uncovers *lurking variables* (Joiner 1981). These are variables that are not in the dataset (or may have been excluded from the analysis), but should be included because they account for important aspects of the data. For example, when a simple model describes most of the data well, an attempt to discover why certain data points fail to conform may turn up a lurking variable that accounts for the unconventional behavior or the outliers. The recognition that the data fall into two or three homogeneous subgroups raises the question of how to characterize those subgroups.

4. AN EXAMPLE: SALARIES OF MAJOR LEAGUE BASEBALL PLAYERS

To illustrate a variety of ideas in exploratory data analysis, we will find it convenient to have at hand a substantial example. For a statistical graphics exposition at the 1988 Annual Statistical Meetings, the Section on Statistical Graphics of the American Statistical Association assembled data on the 1987 salary, 1986 performance, and career performance of 528 major league baseball players (322 hitters and 206 pitchers). They challenged members to analyze the data and present the results together in a poster session at the meeting, focusing on the question "Are players paid according to their performance?"

Hoaglin and Velleman (1992) discuss the data further and compare the results of the 15 groups that participated in that session. To indicate the content of the data set, Table 1 lists the variables in the hitters file. We return to these data in subsequent parts of this chapter to provide glimpses of the philosophy and processes of data analysis.

5. PRINCIPLES OF EDA

All statistical data analyses work with models or descriptions of the data and their relationship to the world. *Functional models* describe patterns and relationships among variables. *Stochastic models* try to account for randomness and fluctuation in the data in terms of probabilities. Together these two components form the basis for inference. In practice the distinction between the two assumes that the part of the data assigned to the stochastic model contains only random behavior. Exploratory methods examine and refine functional models

for data without depending on assumptions about the stochastic model.

Velleman and Hoaglin (1981) and Hoaglin, Mosteller, and Tukey (1983) identify four underlying principles of EDA: Revelation, Residuals, Re-expression, and Resistance.

Revelation emphasizes data display to reveal patterns and relationships. EDA starts from the premise that we probably do not know all that is in our data. There may be surprises, errors, extraordinary observations, or unexpected patterns. Well-designed data-analysis displays reveal the unexpected gracefully, but appropriate calculations can provide revealing numbers as well as the input for displays.

Residuals are what remain after a summary or fitted model has been subtracted out of the data. We write

$$residual = data - model$$

to represent this operation. An exploratory analysis takes the data apart step by step, as one might peel an onion one layer at a time. At each step we refine the model and use the residuals to see deeper.

Of course, statisticians have long recommended examining residuals. But the examination has traditionally been a supplementary check for gross violations of assumptions. By contrast, exploratory data analysis views the examination of residuals as the beginning of the next step in the process and anticipates that it will often yield new insights or provoke new questions.

Re-expression involves finding a scale or transformation of each variable that simplifies the analysis of the data. As Mosteller and Tukey point out (1977, p. 89),

> Numbers are primarily recorded or reported in a form that reflects habit or convenience rather than suitability for analysis. As a result, we often need to re-express data before analyzing it.

Hoaglin (1988) discusses several examples from everyday experience—such as the Richter scale for earthquakes, the decibel scale for intensity of sounds, average speed in auto races, and gauges of shotguns—in which the data are already in a transformed scale by the time we hear about them.

Exploratory data analysis emphasizes the benefits of re-expressing data early in the analysis. Re-expression to a

1986	Career
At bats	At bats
Hits	Hits
Home runs	Home runs
Runs scored	Runs scored
Runs batted in	Runs batted in
Walks	Walks
	Years in the major leagues
Put outs	Player's name
Assists	
Errors	
Salary (thousands of dollars) on opening day 1987	
League at end of 1986	
Division at end of 1986	
Team at end of 1986	
Position in 1986	
League at beginning of 1987	
Team at beginning of 1987	

Table 1: Variables for hitters in the baseball salary data.

scale such as the logarithm, square root, or reciprocal can simplify patterns and relationships in several important ways. Most often we choose transformations to:

- improve additivity of a response in relation to two or more factors,

- straighten nonlinear relationships,

- promote constant variability across groups or constant variance of measurement across levels (or over time), and

- make univariate distributions more nearly symmetric.

Fortunately, these benefits tend to occur together; a re-expression chosen for one reason often helps to improve the data with respect to the other reasons. When we must choose, the earlier items on the list usually dominate the later ones, but sometimes only the later purposes can be achieved.

Analyses and displays after careful re-expression are often simpler and more likely to reveal both patterns and unexpected deviations from these patterns. Analyses that were complex and confusing may become simple and straightforward when the data are re-expressed appropriately.

The broad value of flexible re-expression is one of the most effective and enduring lessons of EDA. Traditionally, re-expression has been treated either as an awkward *ad hoc* technique or as another parameter (for example, an exponent) in need of estimation. EDA gives re-expression a central role in developing functional models and offers ways of finding re-expressions simply. The re-expressed variables usually lead to more straightforward and more easily understood functional models.

The baseball salary data offer a good example (Figure 1). An initial look at the dependent variable, *salary*, reveals that these data are skewed, but re-expression in the log scale does much to improve symmetry. Higher on the list of benefits, the re-expression helps to straighten the relationship between *salary* and virtually every one of the potential explanatory variables. Indeed, among the data analysts that participated in the poster session, those who did not take logs of *salary* generally were less successful.

Once we have re-expressed *salary*, a scatterplot of log(*salary*) against *years in the major leagues* (Figure 2) shows a pattern that we can summarize in three pieces: *salary* is level for about 2 years, grows linearly up to 7 years, and then levels off beyond 7 years. Such a description seems quite plausible, but it is not what we might have expected.

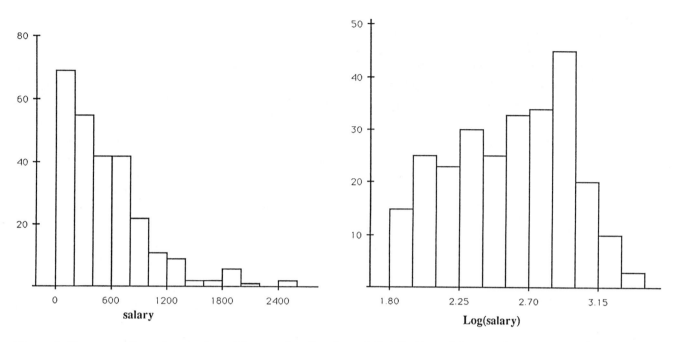

Figure 1: Re-expressing *salary* by logarithms makes the skewed distribution of the raw data more nearly symmetric.

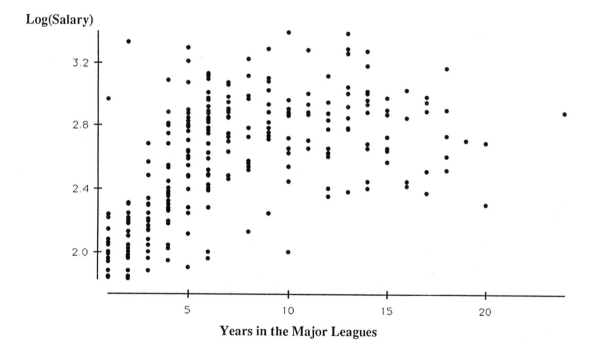

Figure 2: A scatterplot of log(*salary*) against *years in the major leagues* shows a nonlinear pattern.

Resistance protects analyses from the effects of extraordinary values in the data. In classical EDA, resistance comes from using methods that protect automatically against such outliers. For example, medians replace means because they are affected only slightly by occasional extreme values. Order-based statistics such as the median and quartiles appear in boxplots, and in fitting methods such as the resistant line and median polish.

Although resistant and robust methods can provide additional protection in computer-based analyses, they have not been widely accepted. A more popular alternative approach uses traditional methods (perhaps after re-expression) but remains alert to the possible effects of outliers or separate subgroups of observations that follow different patterns. Analyses must then provide ways to diagnose, identify, and isolate outliers and subgroups. Computer-based EDA often relies on careful *recognition* of extraordinary data values in data displays, followed by setting them aside for separate analysis, and on traditional methods rather than on methods that have resistance built in. Thus exploratory data analyses can result from the exploratory use of traditional methods.

For example, a careful look at Figure 2 shows two players with only a year or two of experience being paid surprisingly large salaries. We question the values first because they do not fit the overall pattern established by the other observations, and second because it seems unlikely that rookies would be paid so well.

Recognizing outliers and setting them aside or correcting them draw these points to our attention, which is consistent with the EDA philosophy. Outliers should never be simply omitted and forgotten. We identify outliers specifically so that we can examine them with extra care. Recognizing subgroups in the data and analyzing each group separately is similarly consistent with EDA. The EDA philosophy makes no commitment to a particular model for the data, nor to forcing all the data to fit a single model.

Any competent data analyst would omit an obviously erroneous data value. EDA, however, maintains that it is usually best to resist the effects of outliers on a fitted model, *even if we do not understand why they are extreme.* EDA differs from many other data analysis philosophies in advocating resistance as a fundamental principle. In effect, the exploratory data analyst asserts a belief that simple functional models that fit most of the data well are likely to be more useful for both prediction and understanding than complex models that try to fit all of the data and usually fit less well.

Often data points identified as outliers (and omitted from the model-fitting procedure) are ultimately found to be in error or perturbed by external influences. Nevertheless, EDA neither predicts that this will happen, nor requires that it be so. In the case of the outliers of Figure 2, we can look up the original data source to determine that both of these players have had the wrong value recorded for *years*. In other studies, however, we may have no way to check data values.

EDA maintains that it is usually best to resist the effects of outliers on a fitted model, *even if we do not understand why they are extreme.*

Often we can find univariate outliers more readily when the data are symmetric, and most of the performance variables are more nearly symmetric in the square root scale (a frequent occurrence for counted data). Thus a normal probability plot (see Chapter 8) of the square root of *runs scored in 1986* (Figure 3) shows a tail of suspicious points at the left. All eight of these proved to be errors in the data file.

When we fit a regression model to predict log(*salary*) from some of the performance measures, some further errors in salary figures show up as outliers in the residuals (Figure 4). Among the three labeled points, the salary values for Jeffrey Leonard and Steve Sax were grossly in error (ultimately traced to mistakenly transcribing the adjacent salary in the alphabetical list), but Pete Rose's salary was correct.

On the whole, appropriate use of the general principles and attitudes of EDA contributes much toward building an effective model for the baseball salary data. From further analysis it appears that hitters are indeed paid according to their performance, but that performance is measured in run production rather than in fielding.

6. FUNCTIONAL MODELS

Data analysis fits models to describe patterns in data. Although particular applications may require flexibility in choosing the form of these *functional models*, some models are standard. An understanding of common functional models, ways to display them, and ways to fit numeric descriptions of them provides a good basis for understanding how to look at data.

In the list that follows we emphasize the variety without

√ 1986 Runs

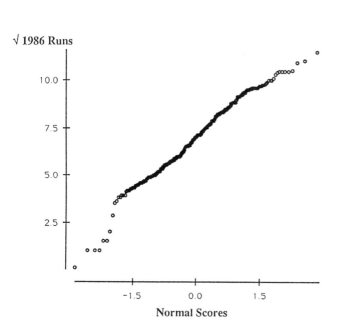

Figure 3: A normal probability plot of the square root of *1986 runs scored* shows that most of the data are nearly Gaussian, but suggests eight observations that seem extraordinary.

residuals

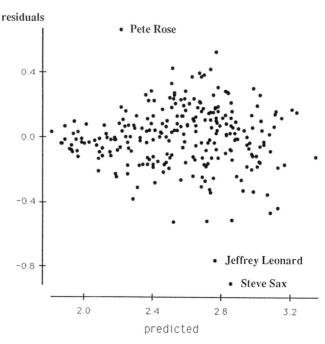

Figure 4: Regression residuals show three outliers. Two of them are errors in *salary*, but one (Pete Rose) is simply extraordinary.

trying to include detailed descriptions. The interested reader can find detailed advice in the following texts:

- For basic pencil-and-paper exploratory data analysis methods, Tukey (1977) and Velleman and Hoaglin (1981).

- For more theoretical discussions of these and more advanced methods, Mosteller and Tukey (1977), and Hoaglin, Mosteller, and Tukey (1983, 1985).

- For regression-based methods, Mosteller and Tukey (1977), Weisberg (1985), and Montgomery and Peck (1992).

- For an overview of multivariate methods, Gnanadesikan (1977).

One-Variable Patterns

In analyzing a single variable we most often consider how the values are distributed. If the data are category names, we are likely to count the members of each category. We might display the relative counts with a bar chart or summarize them with a frequency table. If the categories have a natural order, we would order the bars of the bar chart.

If the data are numeric values, we usually consider how the values are distributed. Stem-and-leaf displays and histograms (especially for larger batches of numbers) are natural displays (as in Figure 1). We typically examine single *versus* multiple modes (a possible indication of inhomogeneity), extreme values (possible outliers), symmetry *versus* skewness (a possible indicator that re-expression would help the analysis), and the spread of the extremes or *tails* of the distribution relative to the spread of the center (often, as compared to the normal distribution).

Most common summary values are simple measures of location and spread. To resist the effects of possible outliers, EDA usually relies on the median and quartiles rather than on the mean and standard deviation, but any similar measures can serve if used with care. Five-number summaries (median, quartiles, and extremes) can point to skewness.

Log(Salary)

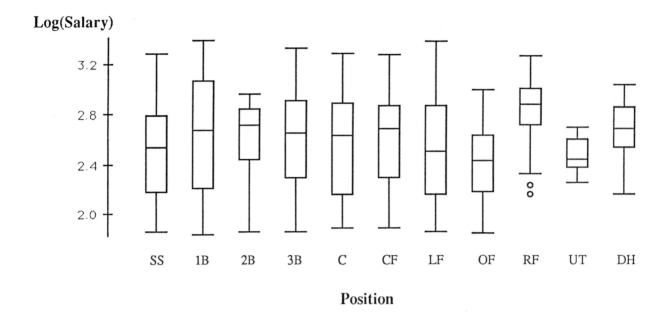

Position

Figure 5: Boxplots of log(*salary*) by position show little difference in level among the positions. Two right fielders with possibly extraordinarily low salaries are displayed separately.

If we have a standard distribution in mind, a quantile-quantile (Q-Q) plot or probability plot, as in Figure 3, is an effective way to assess similarity of distributions. Chapter 8 gives a definition and another example.

Side-by-Side One-Variable Patterns

Often we wish to compare groups by summarizing each, side by side. For example, in the baseball salary data example, Figure 5 shows boxplots of log(*salary*) by fielding position (omitting pitcher). In this EDA display the box covers the middle half of the data (from the lower quartile to the upper quartile), the line across the box shows the median, and a whisker at each end extends to the most extreme observation that is not an outlier (according to a rule of thumb). We can readily see that salary level depends little on position, although designated hitters (DH) and utility fielders (UT) show somewhat less variability. Exploratory displays such as boxplots make it easy to focus on key comparisons of location, scale, and symmetry, while identifying (and limiting the influence of) outliers. A standard confirmatory analysis for this situation is the one-way analysis of variance (ANOVA). It assumes that the variance is equal across all groups and compares the means of the groups.

We typically look for differences or similarities among the groups, identifying any group that stands out from the rest. We also look for overall patterns. For example, if the spread of the data increases as the center increases, it suggests that re-expression (perhaps by square root or log) would simplify further analyses (especially if the individual groups show skewness toward larger values). A linear pattern in the scatterplot of log(*spread*) versus log(*center*) for all of the groups together suggests re-expressing to the power (1 − *slope*) (where a slope of 1 suggests the logarithm).

If the groups are sets of observations of a process taken at time intervals, we should order the plots and summaries accordingly. Now questions of similarity among the groups are equivalent to asking whether the process is *in control* (see, for example, Moore and McCabe 1989, Section 2.1), and we might seek overall measures of location and spread, pooling the individual group spreads as a guide for detecting a drift away from the overall center and thus possibly out of control.

Opportunities for one-variable analysis also arise in comparisons of several measurements taken on the same individuals. For example, a more extensive set of baseball data might give players' performance and salary for each of several years. We could work with each year separately, but we gain by being able to calculate and analyze differences among the years. In Figure 2 the data for a single year suggest that a player's salary will increase during his first seven years. The year-to-year differences would tell us whether such increases actually occur.

Interactive computers can offer ways to enhance our understanding of such data. For example, if a boxplot shows an interesting (possibly outlying) point, selecting it or changing its color can also light it up or change its color wherever else it appears. In this way, we can discover whether the individual is consistently extraordinary or whether one measurement might be suspect.

Two-Variable Patterns

The most common way to visualize two-variable relationships among quantitative variables uses a scatterplot. Typically, we study how the *response y* changes in relation to the *factor x*. We can summarize such a relationship with few assumptions about the data by smoothing the y-values against their x-order. For example, one group analyzing the baseball salary data smoothed the scatterplot of log(*salary*) versus *year* with a robust method known as lowess or loess (Cleveland 1979, 1985). The resulting smooth trace suggested a functional model for how log(*salary*) depends on *year*.

A smooth trace gives a general idea of the level of the data. Smoothing the values above and those below the central smooth produces two additional traces. Together they give some idea of spread as well as level. Here we are interested in the overall pattern of the level—whether it rises or falls consistently, or even linearly. If the rate at which the level changes is itself changing consistently, then a re-expression might straighten the relationship and simplify the analysis. We are also interested in whether the spread remains roughly constant or whether *it* changes consistently. A spread that grows as the level grows is another indication that re-expression might help—especially if the pattern of change in the level suggests a choice of re-expression.

The most common simple function to summarize a scatterplot is a linear trend. Classical statistics uses *least squares*, minimizing the sum of the squared residuals, but other methods are available that are less sensitive to outliers. Typically, the pattern of interest is the slope of the line.

It is especially important to compute and examine residuals when fitting functional forms to scatterplots. The residuals often reveal subtle patterns that were hidden by the gross trend in the relationship. If we find a nonlinear relationship, we might consider re-expressing the y-variable. We should also be alert for outliers or for a serial pattern (in which successive residuals resemble one another more than we would expect by chance). We should also check for patterns in the variability of the residuals—especially a growth of variance with growing fitted values, which might indicate a need for re-expression.

Other functional forms can serve to summarize relationships in scatterplots. Simple polynomials can be used for some non-monotone relationships. Computers have made it practical to fit more general and flexible functions to describe patterns in scatterplots. Methods based on splines (Wahba 1990) and generalized additive models (Hastie and Tibshirani 1990) offer a variety of possibilities. Efron (1988) reviews other approaches that have developed to meet the practical limitations of least-squares fitting.

When only one of the two variables is quantitative, the other variable ordinarily names a group or category. The data then belong to the side-by-side one-variable pattern discussed above.

We might also discover subgroups in the data and consider the side-by-side one-variable patterns comparing these subgroups. This approach is particularly natural when we can identify a lurking variable that accounts for the grouping. For example, we might discover that men and women differ in some important manner not anticipated in the data collection and decide to analyze them separately and compare the results.

Three-Variable Patterns

As the number of variables in the data increases, the variety of structures grows substantially. We illustrate the variety by discussing three leading cases.

The most common three-variable structure summarizes a pair of categorical variables (which distinguish groups or categories). When we have nothing more than two categorical variables, the third "variable" is the count of individuals categorized in each pairwise combination of categories, and the most common display of this information is a contingency table. We most often investigate the independence of the two categorizations. ("Does being female alter the likelihood of being an executive?").

Two categorical variables and one quantitative variable are typically summarized in a two-way layout. Often we ask how the mean of the quantitative *response* changes from row to row or from column to column of the layout defined by the categorical *factors*. For example, we might array average salaries of baseball players according to position played and team to explore the effect of position, of team, and whether the effect of position differs by team.

Fast desktop computers have now made it possible to plot values in three dimensions, revealing the third dimension by making the point cloud appear to rotate.

The most common functional model for a two-way layout is an additive fit that describes each cell mean as

$$cell\ mean\ =\ overall\ level\ +\ row\ effect \\ +\ column\ effect\ +\ residual.$$

This model underlies the classical two-way analysis of variance (see, for example, Hoaglin, Mosteller, and Tukey 1991).

Residuals from an additive fit can be especially informative. Outliers deserve special attention because their impact can propagate throughout the model. The two-way structure also leads to effective ways to assess which re-expressions of the response variable might improve additivity. Resistant methods of fitting are less sensitive to outlying quantitative values. Finally, the residuals can point to the need for an *interaction* term in the model. To fit interaction, we usually must have more than a single observation per cell.

As the third leading case, we may have three quantitative variables. The most common functional model in this case is multiple regression, predicting one of the variables from the other two. Fast desktop computers have now made it possible to plot values in three dimensions, revealing the third dimension by making the point cloud appear to rotate. This is a particularly powerful display because it combines the ability to visualize three-variable relationships with the fundamental operation of multiple-variable summarization, *projection*. Whenever the rotation stops, the static view of a 3-D rotating plot is simply a projection of the points into the plane of the

computer screen. Programs with this capability include Data Desk (Velleman 1989), JMP, and Systat.

Rotating plots offer a new collection of patterns. The simplest pattern is a spherical shape (corresponding in many ways to processes that are in control or scatterplots with a horizontal linear trend). (Rotating plots are most often made with data that have had their mean subtracted and have been divided by their standard deviation, so differences in the scale or center of the individual variables have no effect.) An ellipsoidal cigar shape indicates association among the variables. A rotating plot might have an orientation that reveals distinct clusters of points. Such evidence of inhomogeneity suggests the presence of a lurking variable that defines the groups, and further analysis could explore each cluster individually. A rotating plot that shows a plane suggests a multiple-variable linear fit, such as a regression.

Rotating plots can also reveal patterns beyond the reach of common functional descriptions. For example, two or more "pencils" of points or flat planes strongly suggest inhomogeneity and urge that each group of cases be analyzed separately. (Parallel planes suggest a regression with a discrete "indicator variable.") A cup shape might suggest a quadratic or other more complex fit. A twisted plane suggests that a standard multiple regression will fail to describe the relationship adequately, and may suggest interaction terms or trigger a search for additional variables.

As always in data analysis, identifying a pattern also lets us identify the points that fail to follow that pattern. When the pattern involves three (or more) variables, outlying points can be especially difficult to find, but can be especially informative about the data as well.

Multiple-Variable Patterns

Methods for multiple variables have long relied on matrix algebra, and on presentations that often obscured patterns rather than revealing them. Computers have made multivariate patterns more accessible, if not yet elementary. Although multivariate patterns arise in many situations, they are inherently complex. Simplification inevitably loses some of the richness of the full multivariate structure. A number of methods have been developed for hunting multivariate patterns. Some of these are graphical and rely on the ability of modern desktop computers to animate displays. Again, we discuss two leading cases.

One way to seek patterns among several variables measured on the same individuals is to make all two-variable scatterplots, arrange these plots in a square array according to the variables plotted and *brush* them (Becker and Cleveland 1987). Brushing, like plot rotation, is a computer-based dynamic graphical method. A scatterplot brush is a rectangle on the computer screen controlled by a physical pointing device such as a mouse or pen. As the brush covers points on a scatterplot, they highlight or change color. More important, the points for the same individuals in all other plots on the screen highlight or change color as well. This linkage makes it possible to see whether points lying together in a part of one plot stay together in other plots and, if so, to see where they are and thereby gain a multiple-variable description of them. The brush can be re-shaped to highlight long thin slices on one plot so as to view the other plots "conditionally" on one variable rather than on two.

As always in data analysis, identifying a pattern also lets us identify the points that fail to follow that pattern.

For example, although the performance measures for the baseball players cover a wide range of abilities, many of them tend to go together. A player who produces many hits is also likely to score more runs. Better hitters usually come early in the batting order, where they get more at bats and more opportunities for hits, walks, and rbi's. We can view the extent of these relationships by brushing the array of pairwise scatterplots. In Figure 6 we have re-expressed all the performance measures in the square root scale. The normal probability plots along the diagonal of the matrix indicate that the re-expression has made each variable more nearly symmetric. From the off-diagonal scatterplots, we see that the pairwise relationships among the re-expressed measures are reasonably linear and increasing.

On some desktop statistics programs (such as Data Desk), brushing extends to plots of categorical variables, so that all of the members of a category or group can be highlighted or colored alike in the plot matrix. Brushing can also work with rotating plots—a particularly powerful extension because some orientations of a rotating plot can reveal subgroups whose location in other views might be informative.

Rotating plots themselves can extend beyond three di-

mensions. Although it is difficult (and disconcerting!) to view rotation in more than three dimensions at once, it is reasonable to select three dimensions from a larger space and rotate the plot in those three dimensions, orthogonal to the remaining ones. By selecting dimensions successively, one can explore even a five- or six-dimensional space. Some automated methods for exploring high-dimensional space have been proposed as well. *Grand tour* methods (Asimov 1985) rotate points through orientations in four or more dimensions in a way that is not disconcerting to the viewer but passes "near" every possible orientation of the data in a moderate amount of time.

By far the most common functional model for multiple variables is linear regression, which describes the response variable as a linear combination of suitably chosen functions of other variables. The past two decades have seen the development of a variety of diagnostic tools to help identify individual observations or groups of observations that might exert an undue influence on the regression or that fail to fit the regression. Any data analysis using regression should make extensive use of these methods (see Chapter 8).

For example, most of the groups analyzing the baseball salary data chose multiple regression. Figure 7 shows a successful regression that reflects many of the steps discussed here. *Salary* has been re-expressed by logarithms, *career runs* has been transformed by dividing by *years*, and the piecewise linear pattern noted in Figure 2 has been described with the variable *3lineyrs*, which is 2 for the first two years, *years* for the next five years, and 7 thereafter. Outliers noted in Figures 2, 3, and 4 have been omitted from the analysis.

The regression diagnostics for the analysis in Figure 7 point out once again that Pete Rose is extraordinary. In 1987 Rose was a playing manager—a very unusual combination and one that made him unusual in these data.

Even simple diagnostic displays can reveal unexpected patterns. Figure 8 shows a scatterplot of the residuals from the regression of Figure 7, with the catchers highlighted. All but seven of the catchers have positive residuals. It appears that catchers may be paid more than we might expect from a model based only on performance and experience. Here again, we have found a subgroup of the data that may be special. We know from Figure 5 that catchers are not paid more as a group than players in other positions, so we might suppose that catchers

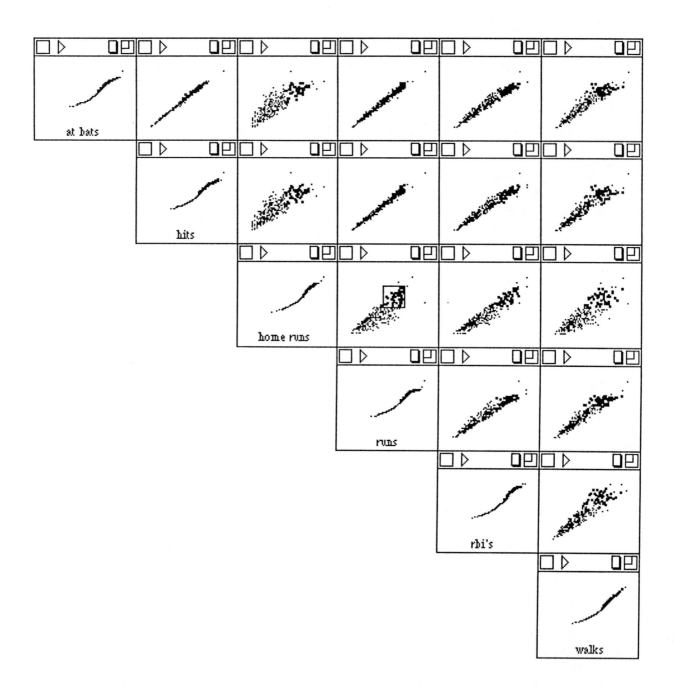

Figure 6: A scatterplot matrix of career performance measures for the baseball players. The diagonal cells show normal probability plots. The points covered by the brush highlight in all of the plots simultaneously.

Dependent variable is: *Log Salary*
322 total cases of which 64 are missing

$R^2 = 81.5\%$ $R^2(\text{adjusted}) = 81.2\%$
s = 0.1658 with $258 - 5 = 253$ degrees of freedom

Source	Sum of Squares	df	Mean Square	F-ratio
Regression	30.5932	4	7.648	278
Residual	6.95694	253	0.027498	

Variable	Coefficient	s.e. of Coeff	t-ratio
Constant	1.35415	0.0395	34.30
yrs	−0.01631	0.0040	−4.06
3lineyrs	0.166627	0.0096	17.40
rnr/yrs	0.006875	0.0008	9.11
run86	0.002579	0.0006	4.25

Figure 7: Least-squares linear regression provides a good fit once appropriate re-expressions are made and outliers are removed.

perform less well in run production than players in other positions, but are not penalized for this in their salaries.

7. PHILOSOPHICAL DIFFERENCES

We may describe the EDA philosophy in greater detail by contrasting it with conventional approaches to data analysis. Although the tenets of EDA philosophy are quite different from those of an idealized conventional statistics philosophy, we need not adhere religiously to one or the other. Each is valuable at different stages of a study, in different ways with different data, and according to the purposes and needs of the analysis. Nevertheless, it is helpful to have a clear idea of the differences and similarities.

Foundations

Role of assumptions. Conventional statistics proceeds as if the data have been collected under the assumptions of inferential statistics. Nonparametric methods have tried to relax some of these assumptions, but they still anticipate a testable hypothesis (which ought to have been specified before the data were collected). EDA can work with such data, but it works just as readily with serendipitous data.

Many other multivariate techniques can be used to ex-plore data. Each provides a different attempt to meet the conflicting challenges of simplifying complex structure enough for the human mind to visualize while preserving enough of the richness and complexity of multiple-variable relationships so that interesting patterns are not lost by the analysis process itself. Gnanadesikan (1977) discusses many methods.

Role of stochastic model. Conventional analyses typically work with a single functional model and pay particular attention to the stochastic model, which must be specified independent of the data. The exploratory approach considers the form of the functional model a matter to be determined both by the purpose of the analysis and by the data. We may use general stochastic models to guide re-expression—for example, when we try to make observed distributions more nearly Gaussian.

The Role of Alternatives

Heterogeneity. Conventional analyses deal with homogeneous populations. EDA encourages us to identify subgroups, to compare and contrast them, and to model each separately if necessary. As a consequence, EDA methods can help to identify subsets of the data that are more suitable for conventional methods. Heterogeneity of the stochastic structure is also of interest to EDA. We would want to identify any subgroup that exhibits higher variability in a key measurement or a skewed distribu-

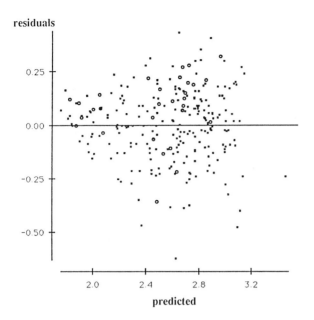

Figure 8: Residuals from the regression of Figure 7 suggest that catchers (plotted with larger points) are paid more than we might have expected from this model.

tion when the bulk of the data is symmetric, although we are likely to need a variable to define the subgroup.

Outliers. Conventional approaches teach that we may omit points from an analysis only when we know that they are wrong. By contrast, EDA teaches that a resistant description of the data often will show us more. Such descriptions find a good fit to most of the data but do not permit extraordinary observations to affect the fit. EDA requires no reason to isolate an observation other than its failure to follow the pattern established by the body of the data. Of course, EDA welcomes an explanation if one can be found. The observations we identify for such special treatment may well be the most informative part of the data and deserve our special attention.

Alternative functional models. Conventional methods expect a single functional model as part of the analysis specification. Though one may perform secondary checks for model adequacy, there is no expectation that the model will change very much. EDA seeks better models and is willing to discard one model in favor of another that fits the data better, or even to carry alternative models forward side by side.

EDA as Scientific rather than Mathematical Statistics

Optimality. Mathematical statistics yields precise answers. If only we can agree on what to assume about the data and about the world, and if those assumptions are true, it is often possible to select an "optimal" procedure. EDA sacrifices such idealized precision in favor of more realistic views of the data. Mosteller and Tukey (1977, Chapter 2) laud the value of analyses that *indicate* a conclusion but can do no more. Indication can often show the way to a more insightful (but less certain) understanding of the data.

EDA teaches us to use hypotheses primarily for guidance; we prefer an approximate answer to the right question over an exact answer to the wrong one.

Hypotheses. Conventional statistical inference requires that we formulate specific hypotheses and test them with formal procedures. By contrast, EDA teaches us to use hypotheses primarily for guidance; we prefer an approximate answer to the right question over an exact answer to the wrong one. Two problems with classical inference from the point of view of EDA are that the hypotheses are often forced to fit the Procrustean bed of the formal procedure and that the path of the analysis is not permitted to stray far from the hypotheses. If we are not careful, we can find ourselves testing absurd null hypotheses or assuming unbelievable things about the data and ignoring or failing to see interesting alternatives. Of course, EDA sacrifices the equanimity that some derive from a "significant" test result—but that equanimity is ill-supported, misplaced, or even treacherous when we cannot trust that the assumptions required by the test were met.

When data are collected with due care (for example, in a well-designed and well-executed experiment), classical analysis is essential and provides "the conclusions" of the study. But even in such instances, a careful exploration can turn up extraordinary observations or unexpected patterns that reveal much about the data and enhance the conclusions of the conventional analysis with important insights.

Practical Data Manipulation

Re-expression. Conventional approaches analyze the data "as is." By contrast, an exploratory analysis often begins by re-expressing the data. Conventional analyses have shunned such preemptive re-expression because the lack of methods for choosing an "optimal" re-expression forces us to use judgment, and because the formal properties of any subsequent analysis are then conditional on the re-expression. The decision to re-express variables and to investigate alternative re-expressions fundamentally changes the character of most analyses and may lead to an entirely different understanding of the data. However, if we accept the EDA argument for re-expression, subsequent conventional analyses of the re-expressed data may be appropriate (conditional on the re-expression.)

Scales of measurement. One school of thought about conventional analyses is especially common in the social sciences (where it is due primarily to the work of S. S. Stevens (1946)). This school identifies measurement scales as nominal, ordinal, interval, and ratio, and holds that there are "...limitations that the scale type places upon the statistics one may sensibly employ." (Luce 1959). EDA holds a diametrically opposite view. Indeed, most common data re-expressions are not permitted by Stevens's rules. EDA uses a wide variety of methods to view and describe data with few assumptions about their structure. Our goal is understanding rather than the imprimatur of significance (see, for example, Wilkinson and Velleman 1992). As Tukey (1961, p. 246) says,

> One reason for the feelings of those who believe that precise scale type should limit the use of statistics may well be the practice, entered into by too many, of regarding statistical procedures as a sanctification and a final stamp of approval. Results based on approximate foundations must be used with the underlying approximation in mind. Those who seek certainty rather than truth will try to avoid this fact. But what knowledge is not ultimately based on some approximation? And what progress has been made, except with the use of such knowledge?

The Role of Data Analysis

Data analysis as a process. Conventional analyses are *end-oriented.* They proceed as if the end (usually a hypothesis test or confidence interval) justifies—indeed, determines—the means. Exploratory data analyses are *progress-oriented.* We learn about both the data and the problem in a series of trials, errors, plots, and data manipulations. Even the decision to end an exploratory analysis is a judgment, for we might always learn more from yet another step of analysis.

Data analysis as part of a larger process. Many conventional analyses are treated as if they are the only study to be considered. Exploratory analyses fit comfortably into the continuum of science, in which it is important to learn how to improve the design of the next experiment or study—in effect, how to ask a better question. This is one important reason for exploring data collected for (and analyzed in) a specific classical analysis.

Although science specifically works this way, the same is true of data analyses for business decisions. EDA is particularly well-suited to analyses of economic and business data intended to help in business decisions because it recognizes these analyses as part of an on-going process.

Exploratory analyses fit comfortably into the continuum of science, in which it is important to learn how to improve the design of the next experiment or study—in effect, how to ask a better question.

Additional data. Conventional approaches limit our attention to the data at hand. Exploratory methods are likely to suggest lurking variables that might not be in the current dataset, but could provide useful additional information.

The Data Analyst as Responsible Participant in the Analysis

Judgment. Objectivity is a stated goal of many conventional analyses, often interpreted to leave no room for human judgment in the midst of a formal analysis. EDA teaches that we should incorporate judgment as a fundamental part of any analysis.

One common demand for judgment is the choice among alternative functional models for the data. Freed from criteria of "best" fit, the exploratory data analyst can judge competing models on their parsimony or intelligibility.

The school of statistical thought known as Bayesian statistics also explicitly employs judgment. However, Bayesian analyses require a specified functional model and usually require, if not a hypothesis to test, at least the specification of a parameter of interest and a region of interesting values. The analyst's judgment about the stochastic structure is then combined with information from the data to reach conclusions about the parameter.

Data mining. Formal statistical inference permits us to answer only the questions originally intended. Many books warn against "data mining" in search of new results. Such a warning is quite correct when applied to classical inference. If we perform many tests, we drastically increase our risk of "finding" something that is not really there; and if we test a pattern's significance because we have seen enough of that pattern to suspect that we can show "significance," then we are very likely to be fooling ourselves.

But in the context of a data-analysis process that uses vague answers and indication to find a way forward, there is less to fear from chance patterns. If we entertain each pattern as no more than a step along the path to deeper understanding, then we need not worry as much whether it is "real." Just as an inadequate scientific theory is exposed by its failure to account for new results, an inadequate data model usually reveals its failings after a few more steps of analysis.

We should not trust the descriptions, patterns, and statistical models we find in the same way that conventional analysts trust their statistically significant results. (But exploration often teaches us to doubt the assumptions that significance requires, so the sacrifice may not be great after all.) To quote Martin Wilk, "The hallmark of good science is that it uses models and 'theory' but never believes them." (Thomas Kuhn might want "ought never" or "never believes them in a crisis" in this statement.) A good data analysis does the same.

Of course, few statisticians blindly hold the views described here. Statistics is practical, and statisticians have often done whatever the data required. There are also alternatives other than those described here. Bayesian philosophy emphasizes the incorporation of individual judgment and eschews testing. Nonparametric methods relax distributional assumptions on the stochastic model. However, Exploratory Data Analysis still stands apart from these approaches in its insistence that the data should speak first.

8. THE ROLE OF COMPUTING

The vast majority of data analyses today are performed on computers. As with many powerful tools, computers exact a payment in return for the power and convenience they offer. It is far too easy to abdicate responsibility to the computer—to "just let the computer do it" (and then, of course, to shift the blame to the computer if the analysis turns out wrong). This temptation is fundamentally contrary to the basic ideas of data analysis.

Traditional batch-processing computer programs running on large mainframe computers may offer much raw power, but they are not suitable platforms for modern data analysis. In *The Design of Everyday Things* Donald Norman (1988) differentiates two ways of getting a task done (p. 184):

> One way is to issue commands to someone else who does the actual work: call this "command mode" or "third-person" interaction. The other way is to do the operations yourself: call this "direct manipulation mode" or "first person" interaction.

Conventional statistics programs are clearly third-person programs. You tell the package what statistics to compute, and it computes them. EDA calls for first-person interaction. Norman's description suits EDA perfectly:

> ...if the job is critical, novel, or ill-specified, or if you do not yet know exactly what is to be done, then you need direct, first-person interaction. Now direct control is essential; an intermediary gets in the way.

Pencil and paper are quintessential first-person tools. Fortunately, computer programs on desktop computers have begun to offer direct first-person interaction.

A data analysis is a journey of discovery. At each step we try to learn more about the data, refine our models for the data, and form new questions using approximate answers to our current questions and informed judgment. The "four R's" guide this journey and define much of what we need from a computer package.

Revelation requires graphs designed for data analysis. These graphs must help us formulate new questions and answer them. For example, the question "Who are the individuals in this group?" can be answered by displaying a name when the user touches a plotted point with a mouse-controlled cursor. Data analysis displays often suggest further analysis steps. An exploratory data analysis program should make it easy to follow these steps. For example, we might ask "How would the analysis change if I omitted this group of points?" and expect the program to show us both the original analysis and the alternative for comparison.

Dynamic data displays reveal much about multivariate data that simply cannot be seen in static displays. Rotating three-dimensional scatterplots and plot brushing can show complex relationships among several variables in ways that reveal unexpected behavior gracefully. As displays suggest further analysis steps, the program should make it easy to take them. For example, if a particular orientation of a rotated plot is informative (perhaps because it shows two separate groups where we expected one homogeneous group), it should be easy to save that orientation so that it can form the basis for further investigation.

Residuals must be available easily and naturally for every fitted model. The first challenge is to make residuals an integral part of the dataset so that they can easily be analyzed and displayed. The second challenge is to provide ways to correct and update the residuals as the analyst changes the conditions of the analysis or the functional model. For example, the question "If I omit this group of points, how will the residuals change?" is a typical exploratory step, as is the question "If I re-express this variable with logarithms, will it remove the apparent bend in the residuals?"

Re-expression must be a basic part of data management and manipulation. Most programs that handle data can perform simple transformations and calculations. For exploratory data analysis, these must be easy to specify, and easy to modify, for often we must seek the best re-expression by trying several in turn. The question "If I change the re-expression of these variables, how will the analysis change?" is a typical exploratory step that should be easy to take.

Resistance in computer-based statistics can come both from displays that help to identify influential data, and from diagnostic methods that help us to be aware of violations of conventional assumptions. Here again, the program must provide the links back to the data as we understand them. Thus, if an observation is dangerously influential in an analysis, it must be easy to identify the observation and to isolate its effects on the analysis.

Regrouping is easiest when displays offer ways to identify subgroups (for example, with different symbols and colors) and tools for specifying and recording subgroups (for example, by circling points on a plot to assign them to a group and creating an indicator (0/1) variable to identify the group).

A data analysis is a journey of discovery. At each step we try to learn more about the data.

Group identity should be clear and consistent in all displays. For example, if males are plotted in blue and females in red, the color scheme should apply to all appropriate plots. Similarly, it should be easy to restrict a particular model or analysis to any subgroup and to compare results across subgroups.

9. GOING BEYOND THE DATA AT HAND

Exploratory data analysis offers powerful tools and a guiding philosophy to help us learn from our data. Nevertheless, we do not have, accompanying EDA, a formal way to draw inferences from the data at hand to the population at large, once we have explored our data and built models to describe them.

The ways in which we use exploratory and confirmatory methods together depend in part on the nature of the investigation.

If the data arise from an experiment that has been designed with suitable care to test a specific hypothesis, then the intended confidence analysis commands our attention. Exploration can serve to reaffirm our faith that the data are appropriately expressed and that no outliers are contaminating the inference. Especially when the analysis will support important decisions and when others will review it, exploration can provide peace of mind. We might also explore in the hope of proposing alternative models and hypotheses for future investigation.

However, if the data are serendipitous, if they were collected (however carefully in randomization and blocking) for some reason other than the study we intend, or if we simply do not know very much about their pedigree, exploration should be our central concern. In this framework, confirmatory ideas play a very different role. We are no longer engaged in deductive reasoning about a specified hypothesis, but rather in inductive scientific reasoning about what we see in the data. In this framework we look to confirmatory ideas to protect us, at least to a reasonable degree, from overfitting the data at hand and from generalizing too broadly to the population at large. These are serious challenges, not easily met.

Many *a priori* reasons justify believing a careful exploratory data analysis more readily than a traditional classical analysis. Most of these (not coincidentally) recall reasons for preferring one scientific theory over another:

- The exploratory analysis will try to identify and isolate anomalous observations and errors, thereby helping the model to fit the (remaining) data better.

- The exploratory analysis is likely to yield a more parsimonious model, both by re-expressing some variables and by omitting anomalous observations. Such models often have a simple, natural interpretation.

- The exploratory analysis is likely to generate worthwhile questions that may go beyond the data at hand. These can guide future research and inform decisions. Some may provide evidence in support of a modified (functional) model, as when additional information explains an extraordinary observation.

- The exploratory analysis will almost always provide intuitively pleasing graphics that reveal both patterns and violations of these patterns and provide a reassuring overview of the data.

Often when a traditional analysis is required to work with all of the data without re-expressing variables and to fit a fixed, previously specified model, the resulting description is much less informative and useful. Although classical theory says that traditional analysis permits inference to the population, exploratory analysis may suggest that the model is wrong, and thus that the inference is questionable.

Inference procedures can provide a check against overfitting. Diaconis (1985) addresses the problem of when to believe a pattern from an exploratory data analysis. The tools of modern data exploration are quite powerful and may find patterns in some data that are peculiar to that particular dataset. We would like to protect ourselves from relying on such artifacts. It might also be

nice to have the additional benefits of formal confirmatory analyses in which confidence statements help us to understand what to expect of other samples from the same population. How, then, can we draw conclusions beyond the data at hand?

Some Specific Proposals

Personal responsibility. One possibility is simply to take the responsibility on our own shoulders and assert that we believe the model. When the exploratory model is parsimonious and fits the data well, when it is simpler and more interpretable than other models, when the observations it excludes make sense as extraordinary points, and when the analysis raises new and interesting questions, we are encouraged to believe it. Nevertheless, we have no assurances. But the honest classical analyst making conventional assumptions has no assurances either.

Although classical theory says that traditional analysis permits inference to the population, exploratory analysis may suggest that the model is wrong, and thus that the inference is questionable.

We would always be wise, as part of such an assertion, to examine the residuals. The appearance of randomness in the residuals lends some hope that the data were an appropriate sample and that the model has accounted for genuine patterns in the data.

Diagnosis. We will also want to *diagnose* the data with respect to any model we fit. In regression, diagnostic measures such as leverage and DFITS (see, for example, Chapter 8 and the books by Belsley et al. 1980 and Chatterjee and Hadi 1988) and displays such as partial regression plots and Q-Q plots of residuals can give us confidence that the model we have fitted was not determined (either in its form or in its coefficients) by a few influential points or by grouping in the data.

Dynamic graphics. A related approach studies the sensitivity of any important fitted coefficients to the decisions made in the course of the analysis. We might be more secure using a model whose parameters were relatively insensitive to the exact re-expression selected, for example. With sufficient computing power we can study many of these sensitivities empirically. For example, modern statistics packages such as LISP STAT (Tierney 1990) and Data Desk (Velleman 1989) offer the ability to link the value of any parameter in a function (for example, an exponent) to a graphical control manipulated by a mouse. We can then slide the power of a variable over a reasonable range and observe the changes in, say, a plot of residuals. With sufficient computing power, the changes can be animated, appearing to adjust smoothly as the power is changed. One great advantage of such real-time sensitivity analyses is that they exhibit the *velocity* of changes as well as their direction and magnitude.

Monte Carlo experiments. We might imagine trying to gain an empirical understanding of the stochastic behavior of exploratory data analyses by automating the fitting of exploratory models and studying the performance of the automatic methods in a Monte Carlo trial. Although some proposed methods can perform important specialized parts of the exploration automatically, it is doubtful that we will see good automated data exploration in the near future.

Jackknife and bootstrap. Jackknife and bootstrap methods (Efron 1982, Efron and Gong 1983) may provide estimates of variances for fitted coefficients. These methods are usually helpful when the model has been determined as a part of the data analysis process, but they generally are theoretically precise only conditional on any re-expressions and on the isolation of extraordinary observations.

Validation and cross-validation. If we hope to go beyond simply assessing the stability of the model or its sensitivity to the choices we have made—to confirming the choice and form of the functional model, the most promising option at present is validation of the fitted model, either on new data or by using separate parts of the original data. As Diaconis (1985, p. 18) says,

> Replication on fresh data, preferably by another group of experimenters, is a mainstay of "the scientific method."

If a similar independent dataset is available, we can gain a degree of confidence in a model's applicability beyond its "home" dataset if the model fits the second dataset well. When this is not possible, we can split our original dataset at random, perform an exploration of one part, and validate by fitting the model to another part. Often we split the dataset in half, but this is not necessary. It is valid and appropriate to explore *each* half of the data and validate on the other, but it is hard to "forget" what

we learned in the first exploration when we set out on the second.

Validation also must be approached with the principles of data exploration in mind. For example, if we are willing to omit outliers (and especially if we did so as part of the initial exploration), then we should be willing to exclude outliers from our assessment of the success of the model in fitting the validation set. Thus, for example, the sum of squared prediction errors may not be an appropriate measure of success. Instead, we may need to use a robust measure of success, or to identify and isolate observations whose prediction errors are large (or, alternatively, observations that can be identified as outliers within the validation set).

Validation requires that we plan ahead. We may need to isolate part of the data before the initial analysis. Validation also requires sufficient data for both exploration and confirmation.

If we use the full range of these methods, we can emerge with a great deal of confidence in the broad applicability of our data analysis. We can present a simple, understandable, parsimonious model, check that it accounts for most of the interesting pattern in the data, provide evidence that the model is relatively insensitive to subjective decisions about re-expressions and functional form (or, alternatively, note exactly where the sensitivities lie), offer bootstrap or jackknife estimates of coefficient variability, and finally provide validation evidence of the general success of the model. Often we will be satisfied with far less.

10. CONCLUSION

Throughout this chapter we have emphasized principles and philosophy of data analysis (especially EDA) and avoided details of techniques. In this way we offer a perspective different from the introduction to data analysis as a collection of fairly elementary techniques that readers may have encountered. Although we have mentioned mainly the most common forms of data, EDA approaches exist for a variety of other situations, and research continues on still others.

An understanding of the basic attitude of EDA is an important foundation for data analysis. Even without convenient access to the best exploratory techniques, a student or practitioner who has learned the benefits of resistance, the effectiveness of re-expression, the value of always looking at suitable residuals, and the gains from effective displays will have a good chance of extracting the greatest amount of useful knowledge from a set of data.

Paul F. Velleman is Associate Professor and Chair of the Department of Economic and Social Statistics at Cornell University. He earned his A.B. degree in mathematics and social science from Dartmouth College and M.S. and Ph.D. degrees in statistics from Princeton University. He is a Fellow of the American Statistical Association and of the American Association for the Advancement of Science. He has published research in data analysis, statistical graphics, and statistical computing, and is the developer of Data Desk(R), a Macintosh-based package for data analysis and graphics.

David C. Hoaglin is a Research Associate in the Department of Statistics at Harvard University and also a Senior Scientist at Abt Associates Inc., an applied social research firm in Cambridge, MA. He received a B.S. in mathematics from Duke University in 1966 and a Ph.D. in statistics from Princeton University in 1971. Prior to his present positions he was on the faculty of the Statistics Department at Harvard from 1970 to 1977. His research interests include data analysis, robustness, statistical computing, statistical graphics, and applications of statistics to policy problems. He is a fellow of the American Statistical Association and the American Association for the Advancement of Science and an elected member of the International Statistical Institute.

REFERENCES

Asimov, D. (1985), "The Grand Tour: A Tool for Viewing Multidimensional Data," *SIAM Journal on Scientific and Statistical Computing*, **6**, 128–143.

Becker, R.A. and Cleveland, W. S. (1987), "Brushing Scatterplots," *Technometrics*, **29**, 127–142.

Belsley, D.A., Kuh, E., and Welsch, R.E. (1980), *Regression Diagnostics: Identifying Influential Data and Sources of Collinearity*, New York: John Wiley.

Chatterjee, S. and Hadi, A. S. (1988), *Sensitivity Analysis in Linear Regression*, New York: John Wiley.

Cleveland, W. S. (1979), "Robust Locally Weighted Regression and Smoothing Scatterplots," *Journal of the American Statistical Association*, **74**, 829–836.

Cleveland, W. S. (1985), *The Elements of Graphing Data*, Monterey, CA: Wadsworth Advanced Books and Software.

Diaconis, P. (1985), "Theories of Data Analysis: From Magical Thinking Through Classical Statistics," in *Exploring Data Tables, Trends, and Shapes*, eds. D. C. Hoaglin, F. Mosteller, and J. W. Tukey, New York: John Wiley, pp. 1–36.

Efron, B. (1982), *The Jackknife, the Bootstrap, and Other Resampling Plans* (CBMS-NSF Regional Conference Series in Applied Mathematics, 38), Philadelphia: SIAM.

—— (1988), "Computer-intensive Methods in Statistical Regression," *SIAM Review*, **30**, 421–449.

Efron, B. and Gong, G. (1983), "A Leisurely Look at the Bootstrap, the Jackknife, and Cross-validation," *The American Statistician*, **37**, 36–48.

Gnanadesikan, R. (1977), *Methods for Statistical Data Analysis of Multivariate Observations*, New York: John Wiley.

Hastie, T. J. and Tibshirani, R. J. (1990), *Generalized Additive Models*, London and New York: Chapman and Hall.

Hoaglin, D. C. (1988), "Transformations in Everyday Experience," *Chance*, **1**, 4, 40–45.

Hoaglin, D. C., Mosteller, F., and Tukey, J. W. (eds.) (1983), *Understanding Robust and Exploratory Data Analysis*, New York: John Wiley.

—— (1985), *Exploring Data Tables, Trends, and Shapes*, New York: John Wiley.

—— (1991), *Fundamentals of Exploratory Analysis of Variance*, New York: John Wiley.

Hoaglin, D. C. and Velleman, P. F. (1992), "A Critical Look at Some Analyses of Major League Baseball Salaries," in *1991 Proceedings of the Section on Statistical Graphics*, Alexandria, VA: American Statistical Association.

Joiner, B. L. (1981), "Lurking Variables: Some Examples," *The American Statistician*, **35**, 227–233.

Jones, L. V. (ed.) (1986), *The Collected Works of John W. Tukey, Volume III: Philosophy and Principles of Data Analysis: 1949–1964*, Monterey, CA: Wadsworth & Brooks/Cole Advanced Books & Software.

Luce, R. D. (1959), "On the Possible Psychophysical Laws," *Psychological Review*, **66**, 81–95.

Montgomery, D. C. and Peck, E. A. (1992), *Introduction to Linear Regression Analysis*, 2nd ed., New York: John Wiley.

Moore, D. S. and McCabe, G. P. (1989), *Introduction to the Practice of Statistics*, New York: W. H. Freeman.

Mosteller, F. and Tukey, J. W. (1977), *Data Analysis and Regression: A Second Course in Statistics*, Reading, MA: Addison-Wesley.

Norman, D. (1988), *The Design of Everyday Things*, New York: Doubleday.

Stevens, S. S. (1946), "On the Theory of Scales of Measurement," *Science*, **103**, 677–680.

Tierney, L. (1990), *LISP STAT: An Object-Oriented Environment for Statistical Computing and Dynamic Graphics*, New York: John Wiley.

Tukey, J. W. (1961), "Data Analysis and Behavioral Science or Learning to Bear the Quantitative Man's Burden by Shunning Badmandments," in *The Collected Works of John W. Tukey, Volume III: Philosophy and Principles of Data Analysis: 1949–1964*, ed. L. V. Jones, Monterey, CA: Wadsworth & Brooks/Cole Advanced Books & Software, 1986, pp. 187–389.

—— (1962), "The Future of Data Analysis," *Annals of Mathematical Statistics*, **33**, 1–67, 812.

—— (1977), *Exploratory Data Analysis*, Reading, MA: Addison-Wesley.

Velleman, P.F. (1989), *Data Desk*, Ithaca, NY: Data Description, Inc.

Velleman, P.F. and Hoaglin, D.C. (1981), *Applications, Basics, and Computing of Exploratory Data Analysis*, Boston: Duxbury Press.

Wahba, G. (1990), *Spline Models for Observational Data* (CBMS-NSF Regional Conference Series in Applied Mathematics, 59), Philadelphia: SIAM.

Weisberg, S. (1985), *Applied Linear Regression*, 2nd ed., New York: John Wiley.

Wilkinson, L. and Velleman, P. F. (1992), "Nominal, Ordinal, Interval, and Ratio Typologies Are Misleading," *The American Statistician*, to appear.

Chapter 3

Computers and Modern Statistics

Ronald A. Thisted
University of Chicago

Paul F. Velleman
Cornell University

1. THE SCIENCE OF STATISTICS

Statistics has developed from two disciplines: the mathematical study of probability and chance events, and the scientific attempt to draw conclusions from data in the face of inevitable error and imprecision. Modern statistics does not simply apply mathematical results to determine the properties of particular statistical methods; it includes a concern for discerning, describing, and confirming patterns and relationships in data.

It has been said that scientists use models but don't believe them. In this sense, statistics resembles science more than it does mathematics. This attitude, with its bent toward scientific method, shapes both how statisticians use computers and how they teach modern statistics. To a statistician a computer is a tool that assists in understanding the real world. There is an aspect of detective work to statistical data analysis; the rewards often come from discovering something about the world and only rarely from the beauty of the tools that make that discovery possible.

How Computers Affect What Statisticians Do

Throughout its century of existence, statistical practice has combined mathematical theory, methodological research, and applications to scientific problems. Each of these areas has benefited and motivated the others. Almost as soon as they became available, computers began to play a significant role in all three areas. Indeed, in a sense, statisticians had been waiting for 70 years for computers to be invented. Methods such as regression analysis were well understood long before it became practical to apply them. As computers became more accessible and more powerful, these methods came into wide use by statisticians and other scientists alike. With this increase in use came increased experience and familiarity, which, in turn, led to clearer understanding of the limitations of the methods. Further advances in computational power then motivated the development of still newer and less limited statistical methods. These advances also have made it possible to harness computers for theoretical research. Computational advances have changed the face of statistical practice by transforming what we do and by challenging how we think about scientific problems.

Applications

Before the days of inexpensive computers, computational limitations kept most statistical data analyses to only small amounts of data and relatively simple models. Analyses of even 1,000 observations were considered extraordinary, despite such famous exceptions as Francis Galton's compilation of the heights of parents and children, von Bortkiewicz's data on deaths by horse kick in the Prussian army, and Quetelet's discussion of the distribution of the chest measurements of 5738 Scottish militamen, derived from joint frequency tables of height and chest measurement in eleven regiments that apeared in the *Edinburgh Medical Journal* in 1817. Moreover, investigators had to contend with frequent errors introduced in transcription and calculation. (Velleman and Hoaglin (1981, p. 274) mention a number of errors in Quetelet's tabulation of chest measurements, apparently caused by mistakes in transcription and arithmetic.) Before digital computers, statisticians relied primarily on counts of relative frequency as the raw material for investigating these large data sets. Von Bortkiewicz, for example, simply recorded the number of cavalry soldiers killed by horse kick within each of 14 army corps in each of 20 years.

Today, the very things we count are considerably more complex. For instance, within a few seconds we can find that the first three paragraphs of this chapter contain 327 words of which 183 are distinct (because some words are repeated). Indeed, we use one of those words ("the") 14 times in that short space. Many word processing programs can produce such counts at the touch of a button, and professional writers have come to rely on mundane capabilities such as these.

In preparing this chapter we have tried to write clearly (and we hope that we have done so), but we would hesitate to claim that in the process we have produced literature on a par with the works, say, of William Shakespeare. Quite honestly, our first three paragraphs just don't sound Shakespearean. In part, that is because we are drawing on a very different vocabulary than Shakespeare was, despite the fact that the three of us all write in English. Indeed, almost one fourth (41) of the distinct words in the first three paragraphs were never used by Shakespeare in any of his works. For instance, Shakespeare never employed the word "statistical" or "regression" in his plays or poems. Nor did he use the less technical words "attitude" and "available." There are nine words in those three paragraphs that Shakespeare employed on only one occasion in all his works, and, of the 183 different words in our passage, one of them appears more than 27,000 times in the Shakespearean literature ("the" once again).

Computational advances have changed the face of statistical practice by transforming what we do and by challenging how we think about scientific problems.

The ability to count peculiar objects, such as the number of words in our three paragraphs that Shakespeare used exactly 36 times in his works, comes only through computing technology. (There are two such words: "attempt" and "also.") As a result, it is possible to ask new questions and to think about old questions in new ways. Thus, for instance, we can begin to think in quantitative ways of such concepts as the richness of an author's vocabulary and the degree of overlap between the vocabularies on which two different authors are drawing. This framework leads, in turn, to new ways of describing data (statistical models) and testing conjectures in the light of available data (inference).

Our abilities to do complex data analyses have expanded gradually but inexorably with advances in computing. Analyses can be complex for a variety of reasons: large amounts of data, mathematically challenging models that reflect more realistic understandings of the world, or many variables that are related to one another in complicated ways.

Now even desktop computers deal with massive amounts of data. Datasets of even hundreds of thousands of values are manageable with the latest computer workstations. For example, data from the 1990 U.S. census will be published on compact disks, making them (at least theoretically) accessible on a desktop computer. Were these data published only in a printed format, as they were formerly, we could answer only the simplest of questions using them because the burden of extracting relevant information would be too high. Earlier the Census Bureau made census data available to researchers on multi-reel sets of computer tapes. Large computer centers established census tape libraries on mainframe computers, providing access for more researchers. As massive collections of data become cheaper and more easily accessible through advances in computing, statisticians are challenged to develop methods that will make good use of such resources by extracting information from them cheaply using methods that can be understood and applied widely.

New methods of data analysis make it possible to deal with models that, though more realistic, are orders of magnitude more difficult to compute than the relatively simple methods that populate statistics texts. For example, we usually fit generalized linear models (which include such commonly applied methods as logistic regression, loglinear models, and Poisson regression) using an iterative algorithm, each step of which requires an entire multiple regression.

Another example in which the only realistic models are computationally intensive comes from medical imaging technologies such as computed tomography (CT), magnetic resonance imaging (MRI), and single-positron emission computed tomography (SPECT). These methods rely on statistical algorithms for image reconstruction that can require thousands of multidimensional quadratures per iteration. Scientists couldn't even have considered such algorithms before modern computers made the calculations feasible.

With computers we can now explore complicated relationships. No longer must we assume that relationships are linear, that data are homogeneous, or that the data contain no extraordinary observations (or errors). Instead, computers can enhance our ability to examine the data and fit more realistic models.

The generalized linear models that we mentioned earlier can be thought of as extending multiple linear regression methods to a much wider domain of application. Multiple regression, however, remains a remarkably useful technique for describing the nature and quantifying the strength of relationships; and it has become a mainstay of such diverse disciplines as econometrics, quantitative psychology, and market research. But simple linear combinations of predictor variables often fail to capture important elements of our nonlinear world. A simple idea that extends the scope of multiple regression is the generalized additive model (Hastie and Tibshirani 1990), which uses functions of the form $\sum f_i(X_i)$ for fitted values instead of $\sum \beta_i X_i$, where each function f_i is a nearly arbitrary smooth function. The additive-model paradigm selects these functions f_i so as to maximize the quality of the fit between the response variable being studied and the additive predictor. Compared to the effort required to calculate the regression coefficients $\{\beta_i\}$ by least squares in the usual multiple regression setting, such models require enormous computation to obtain the f_i.

Programs that implement graphical methods make it possible to integrate graphical exploration, modeling, and description into the standard practice of anyone analyzing data.

Many other new methods would not exist were it not for the now common use of computers in data analysis. Among the most prominent are diagnostic statistics, dynamic graphics, bootstrap and jackknife, and Bayesian data analysis. Efron and Tibshirani (1991) give an accessible exposition of the ideas and application of the bootstrap, generalized additive models, and classification and regression trees.

Traditional statistical methods are simple and often unrealistic. This is not necessarily bad. Simple descriptions are often the most useful, and there is a tradeoff between simplicity and fidelity to the complexities of the world. It *would* be bad, however, if we had no way of assessing whether the results of our easiest methods might be untrustworthy or inadequate. Data analysts can use *diagnostic statistics* to check for certain deviations from simple structure. For example, even a single extraordinary data value can substantially alter the result of a regression analysis. Diagnostic statistics can pinpoint the errant value and assess its influence on the analysis. Chapter 8 focuses on this topic. Readers interested in learning about regression diagnostics in detail should consult Chatterjee and Hadi (1988) or Cook and Weisberg (1982).

The graphics abilities of modern desktop computers have made it possible to animate statistical graphics. For example, we can depict the relationships among three variables by graphing points in three dimensions and rotating the "point cloud" to display the structure. The low-tech method for doing this is to construct a three-dimensional scatterplot by encasing colored spots in a clear acrylic cube. This cube can then be picked up and examined from any angle. The high-tech approach (which we prefer, for obvious reasons) uses the computer to calculate the projection of this point cloud onto a plane. We can make the point cloud appear to rotate by sequentially projecting from different angles, thereby obtaining the same effects as twisting the acrylic cube. These rotating scatterplots, an example of *dynamic statistical graphics,* are now available as a part of many statistical packages for desktop computers. Other dynamic graphs link several displays so that selecting observations in one causes them to highlight in all simultaneously. For example, see Cleveland and McGill (1988) and Velleman (1989). Programs that implement these graphical methods are now widespread and inexpensive. They make it possible to integrate graphical exploration, modeling, and description into the standard practice of anyone analyzing data.

The examples of statistical methods and applications using them that we have cited illustrate two major roles that computers play in modern statistical practice. First, computers have expanded the collection of tools that one can wield in data analysis, and they have done so in such a way that the analyses become more informative and more realistic. Second, the raw materials with which statistical practitioners work—data—now most often are most conveniently usable in "machine-readable form," that is, as computer files that can be transmitted from one computer to another.

Computers play another key role in the everyday work of those who deal with data. With the advent of word-processing, telecommunications, and graphics programs, as well as statistical programs that until just a few years ago were available only on mainframes, the desktop computer now integrates all the tasks of research. We take notes, do calculations, send mail to our co-workers, share data and manuscripts, and create files and archives—all using the same basic device.

Research

Computers have both raised new research questions and provided new tools for solving theoretical questions in statistics. (Of course, the former have outpaced the latter, making a healthy contribution to the field.)

Applications of statistics have traditionally motivated research. As computers have helped us to grapple with the richness of realistic approaches to data, they have also helped to raise challenging theoretical questions. In addition, the growth of computer power has transmuted some theoretical questions from purely academic interest to great practical importance.

The computer program mentioned earlier that calculated the "Shakespeare counts" was a product of research into the statistical identification of authorship. In 1985, Shakespeare scholar Gary Taylor discovered a nine-stanza poem that he believed to be a genuine (and previously, unrecognized) work of Shakespeare. Taylor noted that the word usage in this poem is sufficiently rich to include words that appear nowhere in the body of accepted Shakespearean works. Because nearly every genuine Shakespearean passage of comparable size contains words that Shakespeare used nowhere else, Taylor took this to be evidence of authenticity.

To a statistician, the number of "new words" found in the disputed text could be greater or smaller than one would expect from the Bard himself. Thus an interesting question is this: Given the known extent of Shakespeare's vocabulary, and the variability he exhibited from one work to the next, is the number of new words in Taylor's discovery too small, too large, or just right? This question can be addressed effectively only with computer power that, until a few years ago, was beyond reach. For an account of Taylor's discovery and a comparison of its vocabulary to that of William Shakespeare, see Thisted and Efron (1987). Donald Foster (1987) gives a compelling argument for why Taylor's discovery is unlikely to be Shakespearean, and Holmes (1985) is a useful review of approaches to quantifying literary style.

Sir Ronald Fisher developed a procedure for testing whether two categorical attributes are independent (when the margins are fixed). The procedure is known as Fisher's Exact Test. It is "exact" in the sense that it is based on combinatorial enumeration of the possibilities, whose probabilities are then summed. The method was feasible only when each attribute had only two categories, because the complexity of the calculations grows exponentially fast with the number of categories. The growth of computer power has rekindled interest in the general case, spawning research into new algorithms for general exact tests. These new algorithms are now finding their way into the catalog of standard statistical methods.

Computers have also become indispensable tools for studying research problems in theoretical statistics. Programs for symbolic algebra, such as Mathematica, Maple, Reduce, and MACSYMA, are becoming as indispensable to statisticians as they already have become to physicists.

For example, while studying the extremal process generated by a marked linear birth process, Bunge encountered the following integral:

$$
I_n = \int_{0 < s_0 < \cdots < s_n < 1} \sum_{i=0}^{n} p^2 (1 - q\, s_i)^{-2}
$$
$$
\times \prod_{\substack{j=0 \\ j \neq i}}^{n} (s_i - s_j)^{-1} \, ds_n \ldots ds_0
$$

for $0 < q \equiv 1 - p < 1$, $n \geq 0$. By using MACSYMA to solve I_n when $n = 0, 1$, and 2, he was able to infer that $I_n = p(-\log p)^n / n!$; and with the answer known, the equality was easy to prove (Bunge and Nagaraja 1991 give the proof of a more general result). Heller (1991) is a useful source for those seeking to learn MACSYMA with statistical problems in mind.

Computer-based simulations make it possible to explore the behavior of new techniques in a variety of settings and to do approximate calculations that are analytically intractable.

Interlude

Computers play several fundamental roles in statistics. They are the tools with which we understand the world, allowing us to use increasingly realistic models and deal with growing masses of data. They motivate research by raising new questions about statistical methods and by providing tools for investigating theoretical questions. It is thus not surprising that we believe that computers also should play a major role in teaching statistics. However, in this area the application of computers has lagged.

2. TEACHING

When we teach statistics as if it were a branch of mathematics, students are left on their own to make the connection between theory and any practical application. Although students of mathematics learn an appreciation for elegant results, there are few mathematicians in basic statistics courses and even fewer elegant theorems to capture their attention. Most of the students in these basic statistics courses are primarily studying other disciplines. They are thus more likely to be motivated by learning about the world, especially the world as seen through the eyes of their own discipline and interests, than by elegant mathematics. They are more likely to have the kind of "Aha!" reaction that makes difficult ideas sink in and stick from an insight about a complex relationship among some real-world variables than from working through the proof of a limit theorem. Students who see statistics as a tool for understanding the world find it empowering.

This view helps us respond to the question in the minds of many students: "Why should I bother with this?" The answer is not "Because it's required for your degree," but rather "Because you can learn more about the world—whatever aspects of it may interest you—by using the tools, methods, and reasoning of statistics."

Students who see statistics as a tool for understanding the world find it empowering. Computers are important in statistics class to the extent that they increase the opportunities for intellectual engagement.

The view of statistics as science also helps to explain the special role of computers in statistical practice and teaching. Computers allow a statistics class to consider the complexities of the real world. We are no longer limited to toy datasets contrived so that all square roots are integers and all quotients have divisors of 10 or 2. Instead, we can show, discuss, and learn from data collected in a study to see whether a new drug can reduce the chance of rejection in kidney transplants. Such real

examples are motivating. Students know that the world is complex. When we oversimplify, we make the subject seem irrelevant.

Computers can make the classroom experience more like real-world statistical practice. Statisticians analyze data to learn something about how the world functions. To do so, they combine knowledge of statistical methods (an unreal world) with knowledge about the subject matter giving rise to the data collection (a somewhat more real world). When we work in class with real data, students, too, can participate in this process, combining the knowledge they are gaining in other disciplines with the methods taught in statistics class. This helps students to see for themselves why they should study statistics, and it also helps them to recognize the limitations of statistical analysis.

Computers are important in statistics class to the extent that they increase the opportunities for intellectual engagement. Students ask questions when the answers interest them and the answers are not hard or impossible to obtain. This is much more powerful and effective motivation than merely seeking to learn the right responses for an exam. Because computers simultaneously are central to what statisticians do and are the tools that make structured inquiry possible, we cannot imagine an up-to-date introductory or applied statistics course that does not give students substantial experience with computers.

Kinds of Statistics Courses

We identify two kinds of statistics courses most often taught to undergraduates. Historically, they have been distinguished from each other principally by their mathematical prerequisites. The first kind includes courses in probability and mathematical statistics, often with a prerequisite of mathematical analysis. Typically, these courses focus on estimation and testing and on classes of statistics that are optimal in some sense under certain mathematically motivated conditions.

The second kind includes both elementary courses and courses in applied statistics. In the past, these courses often have been watered-down versions of the first type of course, intended for students with less mathematical background. Often they, too, have focused on estimation and testing, calling them instead description and inference. However, modern statistical practice seldom gains by separating inference from description. Both "inferential" and "descriptive" methods may apply at almost any stage of a statistical data analysis.

This split based on students' mathematical preparation has meant that those with the strongest mathematical background study statistical theory without much exposure to data analysis or the applications of statistical methods in practice, which are relegated to the mathematically weak.

We believe that both introductory and applied statistics courses should be motivated by the desire to learn about the real world. They thus become courses in the science of data analysis rather than the mathematics (or, formerly, the arithmetic) of statistics. But introductory and applied statistics courses should be quite different from mathematical statistics courses. Mathematical statistics courses should serve students who intend to study the discipline of statistics. Introductory and applied statistics courses should aim at students who intend to do anything else. However, because most students of statistics choose this field because of a broad-based interest in almost everything else, they will usually be interested in both kinds of courses.

Mathematical statistics courses can benefit from a somewhat different computational perspective. Students typically encounter the theory of maximum likelihood, for instance, in the context of exponential-family distributions such as the normal, exponential, gamma, binomial, and Poisson. In these distributions, maximum-likelihood estimators for standard problems have closed-form solutions, and all the calculations of the theory are straightforward. When we omit problems that do not admit closed-form answers, we leave students with the impression that maximum likelihood is useful only when standard assumptions (such as normality of distributions) are met. Worse, they may think that maximum likelihood is merely a technical device for justifying our use of estimators that are "obvious." One need not assume normality to use maximum likelihood, and it is both instructive and broadening to include some less trivial examples in a course on mathematical statistics.

As one such example, consider the situation in which we have a random sample of size n from a population that mixes a fraction, α, of normal with mean μ_1 and standard deviation 1 and $1 - \alpha$ of normal with mean μ_2 and standard deviation 1. We need to estimate the three unknown parameters, μ_1, μ_2, and α. The system of likelihood equations, formed by taking the partial derivatives of the log-likelihood with respect to the parameters, lacks a closed-form solution, but can easily be solved iteratively using one of a variety of computing algorithms. This problem leads naturally to theoretical questions such as, "What is the sampling distribution of $\hat{\alpha}$, and how might we estimate its variance?" A useful source for examples that are both computationally and statistically interesting is Thisted (1988).

Why Use Computers in Teaching at All?

The role that computers should play in teaching statistics depends on the expectations we have of our courses, our students, and ourselves. The particular ways that we (the authors) use computers follow in part from our educational philosophy. There are many different ways to view the *purpose* of a statistics course. It is worthwhile to consider a few of them briefly, because they have very different implications for how—and even whether—computers should be used in statistics teaching.

One view is that courses should equip students for what they might reasonably expect to encounter down the road. This is a market-driven or syllabus-driven approach. In this view, the successful student must be exposed to the right collection of tools that he or she will need to use later. Both teachers and students are externally driven, passive participants following external prescriptions. These prescriptions are likely to reflect the fact that today computers do most statistical calculation, or that employers expect or desire computer competence. The role of computers in this model is to fill a niche in the student's background.

A second view is that statistics courses are a means for transferring knowledge from the professor to the student. In this view, the successful student will know some portion of the facts that we teachers already know and deem to be important. This view places less emphasis on calculation and leaves more time for theorems, philosophical issues, data analysis, and drill. The teacher is active, but the student is still a passive receptor. Knowledge of how to use a statistics package is likely to be on most lists of important skills for statistics, so drills are likely to include the use of such a package—usually one of the packages that are standard on computer mainframes and increasingly available on the desktop.

A third view, and one we find most congenial, is that we are trying to teach students how to approach and make progress on problems they may later face. Here, the successful student will know how to address and solve problems that he or she might encounter in the real world, whether or not they are isomorphic to problems seen in class. This view emphasizes collaboration. The teacher sets an example of clear communication, and the student builds experience. Both parties are active participants. We believe that computing—and particularly desktop computing—can be particularly helpful in achieving the goals of this third view of teaching. Moreover, courses designed with this goal are more likely to convey the philosophy and methods of modern data analysis.

3. COMPUTING AND DATA ANALYSIS

Computers are the fundamental tools of data analysis. Although it is possible to learn a great deal about modest-sized datasets with hand computing and graphing, most analyses use computers. The development of powerful desktop computers has largely resolved the conflict between the need for working directly with data and freedom from the drudgery of calculation.

Graphics

Every science has its theoretical, its experimental, and its observational facets. Statistical graphics makes up the observational side of the science of data. Graphical techniques are the principal tools for identifying patterns, structure, and regularity in data. Because they reveal unexpected behavior in data as readily as anticipated patterns, they are ideal for diagnosing model failures and identifying data points that might be particularly influential or suspiciously extreme. Graphics can be excellent tools for assessing the stability and adequacy of statistical models. At their best, they present intuitively pleasing views of the data that show relationships and patterns in ways suited to human perception. Although this process may sound natural and straightforward, human perception of graphically displayed information is an active area of current research and innovation. See, for instance, Cleveland (1985) and Tufte (1983).

Statistical graphics are also pedagogically attractive. Everyone thinks he knows how to read a graph, so new and unusual graphs don't intimidate students in the same way a new formula often does. Because graphics exhibit data with few prior assumptions, they often encourage new questions. When we analyze data, formulating these questions becomes central to the process through which we come to understand the data and what it is capable of telling us. When we teach, these questions help to engage students in discovering the nature of the real world through statistics.

There are many ways to present information graphically, some better than others. Many statistical graphics fail to meet the standards of readability and intuitiveness.

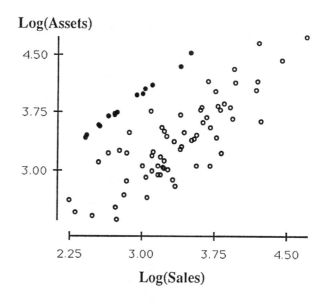

Figure 1: A scatterplot of log(assets) vs. log(sales) for selected Forbes 500 companies reveals a stripe of companies paralleling the rest of the data. It raises a new question about these companies that might not have arisen without the graph.

Most of the charts produced by "business graphics" programs fall into this category, for instance. Fortunately, many simple and useful graphical methods will improve the quality of data analysis and at the same time enrich and enliven classroom learning. Cleveland's elegant discussion in *The Elements of Graphing Data* (1985) is an excellent place for the teacher to start.

Let us consider an example. Figure 1 is a scatterplot of the logarithm of assets vs. the logarithm of sales for 80 companies selected from the Forbes 500 list. The plot itself would be tedious to construct by hand, but only a few clicks of the mouse attached to a desktop computer are needed to produce this figure. We might naturally make such a plot with no expectation of finding any unusual structure in the data. However, in this instance a group of points in the upper left portion of the picture seems to form a stripe, which we have emphasized by darkening the circles that represent each company in the group. Although we hadn't expected to see anything special in the display, now that we see the pattern, it is natural to ask which companies make up that stripe.

When we learn that the stripe comprises all the banks in the dataset, we are then led to still other questions:

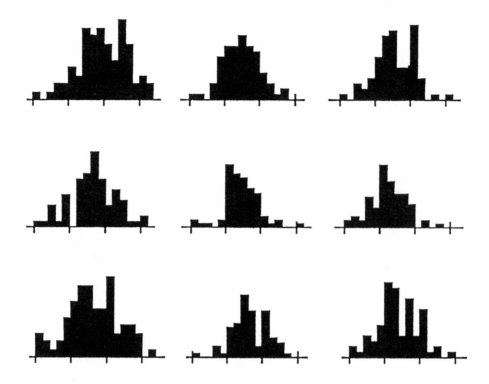

Figure 2: Histograms of nine samples from a unit normal distribution show interesting similarities and differences that help to illustrate the nature of randomness.

"What is different about the banks?", "Are other market sectors similarly identifiable in this plot?"

This example illustrates how a single simple pictorial representation of a set of data can lead to questions about the data that would not otherwise have arisen. Answering these questions in turn leads to a more complete understanding of the data and often leads to improved insights about the world from which the data were collected.

Simulation

Simulation is the "experimental" side of the science of data analysis. In the classroom, simulation experiments play the same role for statistics as laboratory experiments do in many sciences. The teacher can use simulation to illustrate principles or procedures, to demonstrate facts and methods, and to provide a controlled body of experience for the student.

It is helpful to consider the role of laboratory sessions in teaching basic sciences. Typically, experiments in a physics or chemistry lab do not aim primarily to convince the student that the physical or chemical theory is correct; experimental error is a major component in student labs. Nor is the primary purpose to train students in technique—else why bother to compare observation to theory? Well-run science labs offer practical experience in both technique and critical interpretation. Statistical simulation exercises should do the same.

In the classroom, simulation experiments play the same role for statistics as laboratory experiments do in many sciences.

Computer-generated simulations can give students more experience with randomness than they could get in years of practical data analysis. For example, by drawing nine samples of 75 observations from a normal population and comparing their histograms (Figure 2), a student can see that they look different despite coming from the same population. At the same time, the student can notice that they have fundamental similarities despite being independent samples.

Most people find randomness unintuitive and reasoning about it difficult. Students of statistics are no exception. The chance to experiment with randomness improves students' intuition and critical faculties, and it takes some of the mystery out of an aspect of statistics that can be intimidating.

As with chemistry or physics labs, classroom uses of simulation must be carefully designed with a particular lesson in mind, lest students be confused—or even "learn" something incorrect. Many lessons can be taught well with simulation. For example, each student in a class is asked to generate (using a computer) twenty 95% t-intervals with the instruction to notice how many intervals cover the true population mean, μ. This experience with twenty confidence intervals can, with appropriate direction from the teacher, help the student to clarify the difference between population parameters and sample statistics, to note that the endpoints of the intervals are random, and to note that some intervals may fail to contain the true parameter value. Indeed, some students' collections of twenty intervals will have none that miss the true parameter, whereas others' collections will have one, two, or even three or more intervals that "miss." Class discussion of this phenomenon can lead to a better understanding of variability and randomness. It can even serve as an introduction to the binomial distribution, which can be used to describe the variability in this aspect of the students' collections of confidence intervals.

Teachers can also use simulation to illustrate ideas that are otherwise not accessible to beginning students. For example, there are many proofs of the central limit theorem, and many of them are short. But none are particularly intuitive or accessible to students who know nothing of moment generating functions. On the other hand, it is easy to demonstrate the central limit theorem, making the underlying definitions clearer and illustrating methods of simulation in the process.

For example, we can generate 100 samples of 12 observations from a uniform distribution and compute their averages. A simple histogram of these 100 averages (Figure 3) approximates the sampling distribution of the sample average.

Simulations such as this one also help students to clarify the difference between the population parameters (which the student has controlled as part of the simulation) and the statistics (which have been computed). Students can see that each of the 100 sample averages is different,

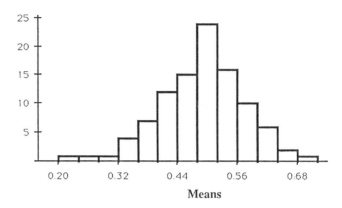

Figure 3: A histogram of 100 averages of samples of 12 drawn from a uniform distribution shows a tendency toward normality.

that none of them exactly equals the "true" value of 0.5, and that they are distributed in a manner that, at least, seems to approach a normal distribution as the sample size grows.

Students gain still more from this simulation experiment if they can discover something for themselves that they might not learn from a classroom demonstration. For example, an exercise might ask them to explore whether the distributional shape of the population (symmetric vs. asymmetric, or unimodal vs. multimodal) affects the sampling distribution of the average. Another exercise might ask them to investigate whether the CLT applies to discrete populations.

Computer-generated simulations can give students more experience with randomness than they could get in years of practical data analysis.

We can use simulation to clarify both the assumptions that underlie standard statistical methods and the consequences of departures from these assumptions. For example, it is easy to generate variables y and x that satisfy the standard assumptions of simple regression. (Generate x as the integers from 1 to 30, generate errors, e, as 30 observations drawn at random from Normal$(0, 1)$, and construct y as $3 + 4x + e$.) Now it is easy to see that the computed regression has neither a slope of 4 nor an intercept of 3, although these are the "true" population parameter values. It is also easy to find out how the regression changes if the underlying relationship is not linear, if the errors are not normal, or if one of the 30 errors is extraordinarily large.

So far, so good. Let's return once again to the chemistry lab for inspiration. Once the student masters lab technique and has some experience under his or her belt, the next kind of lab exercise requires the student to analyze a sequence of "unknowns." Not only must the aspiring chemist draw on his body of experience, he must also develop a strategy, take some blind steps, learn from them, and then converge on a solution.

We can, for example, assign an unknown regression function and a budget for sampling points from this scenario to each of our aspiring statisticians. The student's charge is to report his best description of the function. Along the way, he can discover both the function and any anomalies that might arise in the data giving evidence about the function. In a more elaborate exercise, students might be assessed a cost for each data point observed and offered the chance to specify the x-values at which to "measure" a new y-value. This can provide a motivating introduction to issues of experimental design.

Reality

Computers make it practical to teach with real data rather than with constructed "toy" examples. Real datasets are more interesting, partly because they come with a story. Students see courses that discuss real data as being more relevant. As David Moore notes in his Introduction, the difference between a set of data and a list of numbers is that data are numbers *with a context*. The more students understand about the context of a problem, the more easily they can ask new and relevant questions of the data. Asking new questions encourages them to become engaged in data analysis, and engagement enhances learning.

Real datasets are often too large to analyze conveniently by hand. We need computers to avoid the drudgery of hand calculation. We do not subscribe to the excuse that students need to compute statistics by hand to learn the formulas. Here are some of our reasons:

- The constant repetition of pressing buttons on a hand calculator to compute sums and sums of squares quickly becomes a rote task divorced from the formula.

- Students can work directly with the formulas on computers to learn how the formulas behave and to consider what would happen if they were altered. For example, it is easy to substitute an absolute value for the square and square root in the formula for the standard deviation to see how this changes the result, but it would probably not be worth the effort to do this by hand.

- Once they know the formula, students of applied statistics are unlikely ever again to compute data analyses other than by computer.

Statistics classes already have a reputation for drudgery. We should hesitate to justify that reputation unless there is substantial benefit. Indeed, we must keep this in mind when we design computer exercises as well.

We do not subscribe to the excuse that students need to compute statistics by hand to learn the formulas.

When we configure computers for teaching statistics, we must consider another aspect of reality. Students should work with realistic statistical software packages. It has become fashionable to bundle a statistics program with a textbook. Unfortunately these programs are often highly structured—modular replacements for calculators that give students no feel for the way in which a data analyst would approach data.

By contrast, full-function statistical software packages are open-ended and provide a variety of capabilities. They offer the student a chance to discover something in the data that the professor might not know or to analyze data of interest to the student but not part of the assigned homework. If students are to develop the critical faculties that we hope to teach from laboratory experience, they must have enough options available so that they can take some wrong steps, discover their errors, and learn from them. An adequate software package will not only make these things possible, it must make them easy!

Data analysis is essentially open-ended (see Chapter 2). There is no "right order" in which to explore aspects of a dataset. Data analysis is often nonlinear and involves backtracking. As in good detective novels, observations made early in the game can take on an entirely different color in the light of subsequent discoveries. Also, we may think of several threads to follow, and we can explore only one at a time. It may well make sense to set the current line aside temporarily and go back to develop another. So backtracking and reevaluation are not merely necessary in data analysis, they are essential.

One of the hardest things for students of statistics is that they should ultimately receive little guidance in working their way through a data analysis. Facing a dataset with full freedom to decide how to proceed can be frightening, but students should have this experience first in class, rather than when they must analyze data "for real."

4. OBSTACLES

A variety of difficulties can arise in using computers to teach statistics. It is worth saying a few words about some of the most common among them.

Equipment

The choice of computing equipment often depends on a wide range of considerations, including issues of previously obtained equipment, support, and software. Colleges and universities sometimes have arrangements with particular computer suppliers or manufacturers, and these may limit the range of equipment that can be considered. One brand may be more attractive than another because service is easier to obtain locally, local experts for that brand happen to be available, or influential faculty members have (and are familiar with) that brand.

Among the chief considerations (unfortunately) is *money*. Not only must one consider the initial cost of the equipment, but, to be honest, computer equipment should be depreciated rapidly. In three years the hardware will almost certainly be out of date. In five years it may be so much out of date that new versions of the software will fail to run. Thus even colleges that require incoming students to purchase a computer are in trouble—the first-year student's computer will be obsolete four years later. Even within the useful life of a

particular hardware model, maintenance and repair are inevitable.

Software

The choice of statistics program and overall software environment is now much wider than it used to be. Programs with graphical user interfaces have proved to be easier for students to learn and to work with. Such choices carry implications for selecting both hardware and operating system software. Unfortunately, hardware and operating system choices may be severely constrained by considerations such as those in the preceding paragraphs.

Many excellent statistics packages offer student versions and site licenses, but equipping a student laboratory will still raise concerns about *money*. Software vendors who actively maintain their packages may issue updates annually or even more frequently. The cost comes in three places. First, updates need to be purchased, albeit at a generally lower cost than the original software. Second, someone has to distribute the updated version of the software, either by replacing and checking out diskettes in a lab, updating the copy on a file server, or installing the new program on each of several hard disks. The costs here are in people's time—which could, of course, be purchased with negotiable currency. Third, there is the cost of learning and adopting the changes in the updated software, and incorporating them into lectures, notes, exercises, and handouts. Although it may appear attractive at first, the solution is *not* to purchase software that isn't maintained! Instead, these real costs need to be anticipated and planned for.

Projection

Besides giving students access to computers for their work, many teachers like to lecture with a computer screen projected in the classroom so that the class can see it. Unfortunately, the technology for projecting computer screens has lagged behind the technology for computers and computer graphics. Fortunately, we are seeing advances in this area. Color flat-panel displays for use with overhead projectors are becoming fast enough to keep up with dynamic graphics.

Although these and other technologies are lowering the costs of computer projection, unfortunately the best technology still requires *money*. Projectors also tend to require maintenance, which should be included in budget planning.

Obsolescence

Hardware, software, and projection technologies all suffer from rapid obsolescence. Keep in mind that the "personal computer" was introduced only at the end of the 1970s. The first personal computers capable of running truly effective statistics software did not appear until the mid 1980s. The first statistics packages with graphical interfaces and dynamic graphics appeared in 1986. Flat-screen projectors appeared in 1988. We see no sign that the rate of technological development in any of these areas has slowed. Indeed, we should expect to see it continue to accelerate. All this suggests that we should plan to replace computers, software, and projectors periodically into the next century, all of which costs *money.*

Using computers for teaching and research in statistics carries with it certain costs. These costs are real, and it is not always obvious how they can be met.

Our discussion of obstacles keeps returning to a common refrain—namely, that using computers for teaching and research in statistics carries with it certain costs. These costs are real, and it is not always obvious how they can be met. We mention them not to discourage, but rather to inform. Many institutions, small and large, have found ways to capture the benefits of computing in research, teaching, and data analysis by dealing with these obstacles in ways almost as numerous as the institutions themselves. A helpful guide in thinking about these issues is Eddy's (1986) report of a workshop on the use of computers in statistical research.

Ronald A. Thisted is Associate Professor in the Departments of Statistics, Anesthesia & Critical Care, and the College at the University of Chicago. A graduate of Pomona College in Mathematics and Philosophy, he also has a Ph.D. in Statistics from Stanford University. He is a Fellow of the American Statistical Association. He has written on statistical computing, graphics, and data analysis, and has been an associate editor of several journals of statistics and computing.

Paul F. Velleman is Associate Professor and Chair of the Department of Economic and Social Statistics at Cornell University. He earned his A.B. degree in mathematics and social science from Dartmouth College and M.S. and Ph.D. degrees in statistics from Princeton University. He is a Fellow of the American Statistical Association and of the American Association for the Advancement of Science. He has published research in data analysis, statistical graphics, and statistical computing, and is the developer of Data Desk(R), a Macintosh-based package for data analysis and graphics.

REFERENCES

Bunge, J. A. and Nagaraja, H. N. (1991), "The distributions of certain record statistics from a random number of observations," *Stochastic Processes and Their Applications,* **38**, 167–183.

Chatterjee, S. and Hadi, A. S. (1988), *Sensitivity Analysis in Linear Regression,* New York: John Wiley.

Cleveland, W. S. (1985), *The Elements of Graphing Data,* Monterey, CA: Wadsworth Advanced Books and Software.

Cleveland, W. S. and McGill, M. E. (eds.) (1988), *Dynamic Graphics for Statistics,* Pacific Grove, CA: Wadsworth & Brooks/Cole Advanced Books & Software.

Cook, R. D. and Weisberg, S. (1982), *Residuals and Influence in Regression,* New York and London: Chapman and Hall.

Eddy, W. F. (1986), "Computers in Statistical Research," *Statistical Science,* **1**, 419–437.

Efron, B., and Tibshirani, R. (1991), "Statistical Data Analysis in the Computer Age," *Science,* **253**, 390–395.

Foster, D. (1987), " 'Shall I Die' post mortem: Defining Shakespeare," *Shakespeare Quarterly,* **38**, 58–77.

Galton, F. (1886), "Regression Toward Mediocrity in Hereditary Stature," *Journal of the Anthropological Institute,* **15**, 246–263.

Hastie, T. J. and Tibshirani, R. J. (1990), *Generalized Additive Models,* New York and London: Chapman and Hall.

Heller, B. (1991), *MACSYMA for Statisticians*, New York: John Wiley.

Holmes, D. I. (1985), "The Analysis of Literary Style: A Review," *Journal of the Royal Statistical Society, Series A*, **148**, 328–341.

Quetelet, A. (1846), *Lettres à S.A.R. le Duc Régnant de Saxe-Cobourg et Gotha, sur la Théorie des Probabilités, Appliquée aux Sciences Morales et Politiques*, Brussels: M. Hayez. English translation, 1849, Layton, London.

Thisted, R. A. (1988), *Elements of Statistical Computing: Numerical Computation*, New York and London: Chapman and Hall.

Thisted, R., and Efron, B. (1987), "Did Shakespeare Write a Newly-Discovered Poem?" *Biometrika*, **74**, 445–455.

Tufte, E. R. (1983), *The Visual Display of Quantitative Information*, Cheshire, CT: Graphics Press.

Velleman, P. F. and Hoaglin, D. C. (1981), *Applications, Basics, and Computing of Exploratory Data Analysis*, Boston: Duxbury Press.

Velleman, P. F. (1989), *Data Desk*, Ithaca, NY: Data Description.

von Bortkiewicz, L. (1898), *Das Gesetz der kleinen Zahlen*, Leipzig: Teubner.

Chapter 4

Samples and Surveys

Judith M. Tanur
State University of New York, Stony Brook

1. INTRODUCTION

As the twentieth century nears its close, we live in a world filled with sampling and surveys. Mathematics plays an important role in (among other uses) deriving estimators in sample surveys, understanding their sampling distributions, and estimating their standard errors. This chapter focuses, however, not on the mathematics of sample surveys, but primarily on the basic and important ideas of sampling and on the practical aspects of sample surveys that so often make the difference between success and failure of an actual survey. We need, first of all, some definitions. *Sampling* refers to the choice of individuals from a population by probabilistic methods; a *survey* measures the sampled individuals on variables of interest to the designers of the study. Let us explore some examples of sample surveys.

Unemployment Rates. The United States conducts, as do many industrialized countries, a monthly labor force survey: the Current Population Survey (CPS). Sampling over 60,000 households a month, in the week that includes the 19th, the CPS inquires what each household member was doing (working, looking for work, going to school, keeping house, etc.) during the previous week. From these data, the Bureau of Labor Statistics computes the unemployment rate, reported to Congress and the nation for each month on the first Friday of the following month. (For a more thorough but nontechnical description of the CPS, see Leon and Rones 1989; for a wealth of technical details, see U.S. Bureau of the Census 1978.)

Election Polling. As each election nears, the media present a series of polls predicting the results of the voting. These polls typically use a sample of between 1000 and 2000 "likely voters," and they usually predict the results of the election rather accurately, with a "margin of error" of a few percentage points. From these polls and additional samples ("exit polls") taken outside polling places on Election Day itself, the television networks make their vote projections on Election Night. (See Link 1989)

This chapter focuses not on the mathematics of sample surveys, but primarily on the basic and important ideas of sampling and on the practical aspects of sample surveys.

Societal Trends. Since 1972, the General Social Survey (GSS) has questioned a nationwide sample of approximately 1500 adults each year (with a few gaps caused by funding difficulties) about an enormous variety of attitudes and opinions. Questions range from political affiliation, through opinions on areas of government spending, through job satisfaction, to general happiness and opinions on the appropriate values to instill in children. Most of the questions are asked every year, so the data reveal changes in the attitudes and opinions of the American public over time. In general, the finding from the GSS is that we are very gradually becoming more liberal. These data are made available to social researchers for a very nominal charge. They form the basis for scholarly monographs and research reports and also supply background for newspaper reporters and others in the media. (For a detailed description of the General Social Survey, see Davis and Smith 1988.)

Jury Selection. On a much smaller scale, lawyers frequently commission community surveys of a sample of the population eligible to serve on a jury in the venue from which potential jurors will be drawn. Respondents are typically told something about the issues in the case under litigation and then asked how they would vote if they were serving on a jury for the case. Data from such surveys are used both to shape the arguments being prepared and to aid in jury selection—if people from certain demographic groups tend to be unfavorable to a client's side of the case, a lawyer is well advised to avoid members of such demographic groups as jurors and to use peremptory challenges to dismiss them. (See Press 1989.)

Market Research. Numerous surveys in market research help entrepreneurs assess the demand for products and services and explore consumer preferences for products under development. For example, the Institute for Social Analysis at the State University of New York at Stony Brook has a contract to conduct telephone surveys of people who go to the movies regularly. After hearing a capsule description of a movie in the planning stage, respondents are asked how likely they would be to go see such a movie if it were produced. Such legitimate surveys should be sharply distinguished from the deception in which a telephone solicitor poses as a survey taker in order to make a pitch for a product or to pressure the respondent to buy it.

Each of these surveys samples and questions human beings, and this chapter concentrates on methods for such surveys of human populations. But many of the same techniques used in such surveys apply to sampling other entities. Again, some examples.

Capture/Recapture. To estimate the size of a wildlife population, naturalists and ecologists use the "capture/recapture method." They draw a sample of the population and mark each member of the sample in some way that does no harm. They then release these individuals and some time later draw another sample. The proportion of marked individuals in the second sample provides an estimate of the proportion that the first sample constituted of the population, and hence an estimate of the size of the population. (See Chapman 1989.) The method obviously applies to estimating the size of human homeless populations, and the U.S. Bureau of the Census uses a similar approach to estimate the coverage of the decennial census. About three months after Census Day, the Bureau draws a large sample of census blocks (compact geographic areas that include approximately 300 households) and makes an intensive effort to get basic data—name, age, and ethnic background—on every resident of the sampled blocks. Then they match addresses from the census and this sample to find the proportion of the sample that was enumerated in the census and thence an adjusted estimate of the total population. (See Hamilton 1990; Childers, Diffendal, Hogan, Schenker, and Wolter 1987; and Hogan and Wolter 1988.)

Quality Control. With increasing competition in international trade, U.S. industry is moving more and more to adopt methods of quality control such as those long advocated by W. Edwards Deming (see, for example, Deming 1986) and, in large part under his guidance, long practiced in Japan. One key ingredient of quality control is the sampling of units in the manufacturing process (microchips, ball bearings, car engines, men's inner soles) to test whether they meet specifications, and by inference, whether the process that produces them is operating properly. Similar ideas help to monitor accuracy in the delivery of services. For example, the welfare program Aid to Families with Dependent Children (AFDC) is administered by each of the U.S. states with benefits funded in part by the U.S. government. The program regularly samples files containing records of current beneficiaries. Quality control reviewers then carefully review the eligibility status of each sampled file—often visiting the family's home to verify eligibility. From the amount of benefit dollars found to be mispaid to the sampled cases, the U.S. government estimates the proportion of Federal funds deemed to have been misspent in the state's entire caseload. If that proportion exceeds a specified tolerance, the state is required to repay the misspent funds to the Federal government. (For more details see Kramer 1988.)

Accounting. Sampling also plays a role in accounting. Although the popular image of accountants sees them checking every document for every transaction, in reality they often work with samples. For example, railroads and airlines must divide up fees when freight or passengers travel on several lines during a trip but one line receives the full payment. A sample of tickets or invoices gives accurate guidance on how such revenues should be divided. (See Neter 1989 for more details.)

All of these examples share the aim of generalizing to a larger universe (called the *population*) the findings from a sample. The population may be all non-institutionalized civilians 16 years old or over in the United States, all clients served by AFDC in California in July 1990, or all ball bearings produced by the fifth machine of the Acme Manufacturing Company on February 4, 1991. Regardless, the sample itself holds little intrinsic interest—it merely provides the basis for inference to the population.

This chapter explores the requirements for a survey that make these generalizations possible and reliable. After a section on the basic ideas of sampling, it discusses the various forms of probability sampling. It then considers some special issues in telephone sampling and concludes with a discussion of nonsampling errors.

2. SOME BASIC IDEAS OF SAMPLING

Another facet common to all the examples described above is perhaps less obvious than their purpose to generalize from a sample to a broader population. All use probability methods to draw the sample from the population. *Probability sampling* uses a random device, like the toss of a coin or a table of random numbers, rather than human judgment, to decide which members of the population enter the sample. Technically, every member of the population has a known and nonzero chance of being included in the sample.

Several reasons motivate the use of probability sampling. When a sample is drawn by probability methods, the theory of statistical inference can be brought to bear on the analyses of the results. In particular, the sample itself provides an estimate of the *precision* of the estimates constructed from it. (This is discussed further below.) A sample obtained by non-probability methods may or may not furnish precise estimates of population quantities—but there is, in general, no way of knowing the precision of the estimates it provides.

Further, although the luck of the draw may sometimes produce a probability sample that gives inaccurate estimates of population quantities, the chance of such a "bad" sample can be known and thus can be controlled. No such knowledge of the chance of a "bad" sample is available when non-probability methods are used.

Experience has taught that two kinds of judgment must be ruled out if a sample is to avoid bias—the first is judgment exercised by researchers or interviewers about who should be included in the sample, and the second is judgment exercised by potential respondents about whether to participate in the sample. Use of judgment by researchers or interviewers can introduce *selection bias*; use of judgment by potential respondents leads to *self-selection bias* or *nonresponse*.

No exposition of sampling is complete without a discussion of the *Literary Digest* fiasco, a classic example of the troubles that sampling can get into. After many years of successfully predicting the winner of U.S. Presidential elections, in 1936 *Literary Digest*, using data from 2.4 million respondents, predicted that Alf Landon would win the presidency by a wide margin. This prediction was confounded when Franklin Roosevelt won in a landslide. *Literary Digest* chose names from "every telephone book in the United States, from the rosters of clubs and associations, from city directories, lists of registered voters, classified mail order and occupational data" (*Literary Digest* 1936, p.3) and sent "ballot" postcards to the people chosen.

When a sample is drawn by probability methods, the theory of statistical inference can be brought to bear on the analyses of the results.

Another crucial reason to use a probability sample is the avoidance of *bias*—the results from the sample being systematically so different from what one would get if one examined the entire population that they give poor guidance about the characteristics of the population. Thus, although any particular probability sample may give inaccurate estimates, the freedom from bias guaranteed by probability sampling ensures that on the average (that is, over repeated applications of the method) the results come out right. Non-probability sampling methods have no such unbiasedness property.

Most treatments of the results of the 1936 *Literary Digest* poll attribute the failure to the "unrepresentativeness" of the lists used, an issue of selection bias. Surely in 1936 telephone subscribers and members of clubs and organizations were more affluent than those who had neither phones nor the leisure to belong to clubs. Although the *Literary Digest* had used these same sorts of lists earlier with success, in 1936 the economic policies of Franklin Roosevelt increased differences in voting preferences between the affluent and the less affluent that had earlier been less pronounced. Hence the lists that overrepresented the affluent had been good enough to use for sampling in earlier elections, in which economic status was a less salient consideration for voters, but they were ineffective in 1936.

This interpretation of the *Literary Digest* example illustrates a form of selection bias that occurs when the list or *frame* from which a sample is drawn does not coincide with the population to which one wants to generalize. Other forms of selection bias occur when the investigator uses his/her own judgment instead of a chance mechanism for selecting a sample from a frame. The statistical community itself was largely unaware of the pitfalls of such purposive sampling until the late 1920s. Then Corrado Gini and Luigi Galvani, two prominent Italian statisticians, faced the problem of deciding which few returns from the 1921 Italian census should be preserved while the remainder were discarded. They chose to retain returns from districts that were representative of the country's average on seven variables—and discovered too late that those districts were not representative of the country as a whole on other variables that had not been considered in the choice. (See Gini 1928; Gini and Galvani 1929.)

Probably the most important catalyst for the adoption of probability sampling was a seminal 1934 paper by the statistician Jerzy Neyman that included material he had developed a few years earlier in Poland. (Section 3 of this chapter discusses these concepts further.) Neyman demonstrated the superiority of probability-based sample selection procedures over purposive methods based on assumed forms of regression between control variables and the variable for which estimates of population parameters are desired. Neyman applied earlier work of R.A. Fisher (1930, 1932, 1933) to develop an inferential approach to sample survey data using concepts of repeatable samples; in this paper he labeled the approach "confidence intervals." He showed that only probability-based methods (for drawing large samples) could yield confidence intervals with levels of confidence that are both prescribed in advance and not dependent on the unknown distribution of the population. In this same paper, Neyman also introduced notions of stratification, optimal allocation, and clustering; and perhaps more importantly, he advanced the idea of sampling human populations by defining and choosing primary sampling units much larger than individuals (e.g., counties, precincts), thus laying the groundwork for multistage area probability sampling as we know it today.

In the early days of public opinion polling and market research, a non-probability sampling method called *quota sampling* was often used. Interviewers were given quotas to fill however they wished. For example, an interviewer might be told to find ten respondents, half male and half female. One of each gender should be in his/her 20s, one in his/her 30s, etc. One of each gender should

have income under $20,000, one between $20,000 and $30,000, etc. But within the groups defined by the quotas, interviewers were free to choose their own respondents. Such a procedure generates at least two classes of problems. First, because no rules guide choice of respondents, there is no way to replicate the sample with an expectation of getting comparable results—and we have seen that statistical inferences from the results of surveys presuppose the ability to replicate the sample, at least conceptually. Second, because the interviewers are free to choose their own respondents within the quotas, they are likely to choose the most accessible and the most respectable looking of the individuals who fit the specifications. Thus the people chosen as respondents are likely to differ from those not chosen in systematic ways, and the differences introduce bias into the sample.

Having said that non-probability samples are to be avoided whenever possible, we must note that some situations do not permit probability-based sampling.

Having said that non-probability samples are to be avoided whenever possible, we must note that some situations do not permit probability-based sampling—for example, if only some documents from a historic period survive, or only those members of the population who speak a certain language can be interviewed. Stuart (1968) suggests the device of embedding such a non-probability sample in a higher-order probability structure—for example, by having n interviewers each interview a random nth of the available respondents. The results obtained from each interviewer can be considered an independent replicate of the sample, and inference procedures can be applied. (Such replication techniques are also useful in probability samples to examine the effects of different interviewers or to provide measures of sampling error under a complicated design. Mahalanobis (1944) used this idea frequently in surveys in India.) Of course, in a non-probability setting, if the part of the population that cannot be interviewed is systematically different from the part that can be, then bias in generalizing to the entire population is inevitable. The advantage of the higher order probability structure is simply that it justifies statistical inference to the sampled population.

Let us return to the *Literary Digest* example and note

that, besides the lists of telephone subscribers and members of clubs, *Literary Digest* also used voter registration lists to select their sample—and those registered to vote are exactly the ones who elected Roosevelt in a landslide. So something else besides selection bias must have gone wrong. Indeed, the *Literary Digest* sent out ten million postcard ballots, but only about 25% were returned. Potential respondents could choose whether to participate and could easily opt out. A moment's search of the reader's own memory will probably confirm that one is more likely to respond to such a mail ballot if one feels strongly about the issue. It was those who objected to Roosevelt's innovative and drastic economic programs vehemently enough to vote against him in November who felt strongly enough to return the *Literary Digest*'s ballots earlier in the year. This was a problem of self-selection. (See Bryson 1976.)

This particular kind of self-selection, *nonresponse*, haunts even the best-run sample survey of a human population, and it is usually more severe in mail surveys than in surveys done by telephone or in person. Below we discuss some methods that have been developed for reducing nonresponse and coping with it when it does occur, so that it does not invalidate the results of a well-drawn probability sample.

Other kinds of self-selection are more damaging to a survey than these issues of nonresponse. In some so-called surveys the respondents are purely volunteers. For example, television stations sometimes suggest that viewers dial one number to register a pro vote on an issue under discussion and another number to register a con vote. In such "polls," of course, only those who feel strongly in either direction will call in, especially if the call incurs a charge. And then there is no possibility of telling who might have called but did not (as there is in the *Literary Digest* example and in other cases of nonresponse) and hence no way of telling how far off from truth the results of the "survey" are. A similar example occurs in the book *Women and Love* by Shere Hite (1987). Ms. Hite gave copies of her survey questionnaire to all who requested them, but she also sent them broadside to church groups and other clubs, asking that they be distributed to any members who cared to fill them out. Some 4.5% of the questionnaires thus sent out were returned. Again, we cannot know who chose not to respond, and hence we cannot generalize from Ms. Hite's "sample" to the population of American women. Ms. Hite's efforts to do so are misguided, and her attempt to impart a scientific aura to her work by

defending her sampling methods obscures the insights that can be gained from an informal discussion of "what some women out there think."

3. METHODS OF PROBABILITY SAMPLING

As we have noted, a probability sampling method requires that every unit in the population have a known, nonzero chance of being included in the sample. Then, in principle, the sampling procedure can be applied repeatedly (in practice, of course, it is done only once), and such repeated sampling forms the theoretical basis for statistical inferences as propounded by Neyman (1934). Units that have no chance of being sampled (e.g., because they have been omitted from the frame) force the sampler to narrow the inferences. Instead of being able to generalize to the full *target population*, the sampler can actually make statistical inferences only to the *sampled population*.

The simplest kind of probability sampling is *simple random sampling* (SRS). SRS gives every unit in the frame an equal chance of being in the sample, and it also has the property that all samples of a given size are equally probable. Some probability samples have the first of these properties but not the second. For example, assume one wished to choose a sample of size 100 from a frame consisting of 1000 units. One could choose a random digit between 0 and 9 and include in the sample all units whose serial numbers end in that digit. Then every unit in the frame would have a 10% chance of being in the sample. Further, the ten potential samples consisting of 100 units with serial numbers ending in the same digit would each have a 10% chance of being the chosen sample. But potential samples consisting of units whose serial numbers differ in their last digit—such as the sample that includes units 001, 012, 013, 034—would have no chance at all of being chosen. (Such systematic random sampling is discussed further below.)

SRS is equivalent to writing the name of each unit in the frame on a slip of paper, mixing the slips of paper very well, and then drawing a sample containing the required number of slips of paper. The crucial step in this procedure is the mixing, and ample evidence indicates that physical mixing, although often dramatic and convincing, is exceedingly difficult and usually unsuccessful, as is illustrated by the imperfect mixing in the 1970 draft lottery (see Fienberg 1971.) Hence tables of random numbers (e.g., The RAND Corporation 1955) are used to choose a sample; each unit in the frame is given a serial number, and the sample consists of those units with serial numbers corresponding to the numbers drawn from the table.

Most introductory statistics courses concentrate on simple random sampling and assume that the population size is much larger than the sample size. Although this simplification does make for easier exposition, teachers and students should understand that most survey data arise from sample designs more complicated than simple random sampling and that finite populations are common in practice.

Inference from an SRS is straightforward. Let us assume that we are interested in the average income in the U.S. population, μ, that we can draw an SRS of 1600 individuals, that every individual chosen for the sample responds, and that all the respondents report their income accurately and honestly. (Later in this chapter we discuss how to cope with the fact that these latter two assumptions are often violated.) We know that the distribution of incomes in the population is highly skewed—many poor and middle-income people and a few very rich ones—but that fact need not trouble us. The central limit theorem tells us that, if we draw repeated samples of the same size

$$(y_{11}, \ldots, y_{1n}), \ (y_{21}, \ldots, y_{2n}), \ \ldots$$

from this population, calculate the average $(\overline{y})$ of the incomes of the individuals in each sample, and form the distribution of these averages, the resulting *sampling distribution of the sample mean* will be approximately normal. Moreover, the mean of that sampling distribution, $\mu_{\overline{y}}$, equals the population mean, μ. Thus we can use $\overline{y}$ as a point estimator of μ. Further, the standard deviation of the sampling distribution is

$$\sigma_{\overline{y}} = \frac{\sigma}{\sqrt{n}} \sqrt{1 - \frac{n}{N}},$$

where n is the size of the sample, N is the size of the population, and σ is the standard deviation of the original population. The standard deviation of the sampling distribution of a sample mean is known as the *standard error of the mean*.

Several facts about the formula for the standard error of the mean are worth noting. First, $\sigma_{\overline{y}}$ decreases with $\sqrt{n}$; larger samples are more likely to have sample means closer to the population mean than are smaller samples. But because the standard error decreases with $\sqrt{n}$ rather than with n, shrinking $\sigma_{\overline{y}}$ is expensive in terms of sample size—we must quadruple our sample in order to halve $\sigma_{\overline{y}}$. Second, unless the sample constitutes a large fraction of the population, $\sigma_{\overline{y}}$ depends strongly on the size of the sample and hardly at all on the size of the population.

Hence, for a "large" sample, a particular sample mean, $\overline{y}$, can be considered an observation from a normal distribution with known standard deviation. We can then make probability statements based on the normal distribution, from which we can derive, for example, the usual 95% confidence interval,

$$\overline{y} \pm 1.96 \frac{\sigma}{\sqrt{n}} \sqrt{1 - \frac{n}{N}}.$$

(Of course, 1.96 and 95% are arbitrarily chosen for this example; other levels of confidence and corresponding critical values are possible.)

In the above formulation we set up an elaborate mechanism to draw a sample, make a point estimate of a population mean, and place a confidence interval around that point estimate. We tacitly assume that we know σ, the population standard deviation. In practice we would seldom know σ when μ is unknown; nevertheless we need to know σ in order to derive the standard error of the mean and to construct confidence intervals based on the normal distribution. When σ is unknown, it is customary to use the sample standard deviation,

$$s = \sqrt{\frac{\sum(y_i - \overline{y})^2}{n-1}},$$

as an estimate of σ. Then we base confidence intervals and other inferences on Student's t distribution with $n-1$ degrees of freedom. For the sample sizes usual in surveys, however, Student's t distributions are practically indistinguishable from the normal distribution.

As the most easily understood method of sampling, SRS often serves as a touchstone to evaluate other methods of sampling. But it has drawbacks. Because simple random samples are sometimes difficult and expensive to draw, samplers often use two convenient variations on SRS instead: *systematic random sampling* and *probability sampling with quotas* (Sudman 1967). Further, when an investigator needs to make comparisons among subgroups of the population, an SRS may not ensure sufficient coverage of some subgroups to support inferences. Such an investigator can turn to *stratified sampling*, which often has the additional advantage of reducing the standard error of $\overline{y}$. Still further, SRS takes no account of the geographic location of the units chosen for the sample. If interviewing is to be done by telephone, the geographical scattering of the units does not matter; but if interviewers must visit households in person, then scattering incurs large costs for travel. The device of *cluster sampling* reduces such costs. Finally, SRS requires a complete listing of the population in the frame, but for many populations, such as the U.S. adult population, no listing exists. *Multistage area probability sampling*, which combines stratification and clustering, reduces the need for such listings to a fraction of the population. Further discussion of each of these sampling methods follows.

Systematic random sampling uses a sampling interval, $i = N/n$, rounded down to an integer for convenience. Then a single random number, $r \leq i$, is drawn; the sample consists of units with serial numbers $r, r+i, r+2i, \ldots$. This method entails less effort than drawing an SRS; and, unless the ordering of the frame involves periodicities (a very rare situation in practice, as Sudman 1983 points out), it gives results comparable to SRS. The sampling for quality control reviews in the Aid to Families with Dependent Children program described in Section 1 uses systematic random sampling in most states.

Old-fashioned quota sampling permitted interviewers to choose their own respondents as long as they met quotas for each sex, age group, race, and so on. No probability mechanism was used. Professional survey takers currently use *probability sampling with quotas*, with multistage area probability sampling down to the block level and then controls on such variables as gender, age, and employment status. This technique often produces usable results: when researchers experimentally split the sample for the 1975 and 1976 General Social Surveys (described in Section 1) between true probability methods and probability sampling with quotas, they found no differences between the two techniques other than a deficit of one- or two-person households in the quota samples (Stephenson 1978).

Stratified sampling divides the population into the subgroups of interest or into other homogeneous groupings (on the basis of some variable that is easily measured a priori and that is thought to be related to the major variable under consideration). Thus, if we were trying to estimate the average income of the U.S. population,

we might stratify by race, by marital status, or by geographic area. We would then draw an independent probability sample from each of the specified strata.

If N is the total number of units in the population and N_j is the number of units in the jth stratum, we can define $w_j = N_j/N$ as the proportion of the population in the jth stratum. Then we can estimate the population mean by the weighted mean of the stratum means $\overline{y}_j$,

$$\overline{y}_{\text{strat}} = \sum w_j \overline{y}_j.$$

Similarly, the standard error of y_{strat} is given by

$$\sqrt{\sum w_j^2 \, \text{var}(\overline{y}_j)}.$$

Note that $\text{var}(\overline{y}_j)$ is computed using deviations around the stratum mean—in analysis-of-variance terms, it gives the within-stratum component of variance. The between-strata component of variance (that is, the deviations of the stratum means around the overall sample mean), which contributes to the variance of $\overline{y}$ in SRS does not enter into the variance of $\overline{y}_{\text{strat}}$. Hence, if the strata are very different from one another, the standard error attached to a mean calculated from a stratified sample may be much smaller than the corresponding standard error for an SRS.

Statisticians have proposed many schemes to help investigators decide the size of the sample (n_j) to be drawn from each stratum. The simplest is proportional allocation, where $n_j/N_j = f$ for all j, and we can speak of a sampling fraction of $f = .1$ or $.001$, say, across strata. Such a sample is called *self-weighting*, because we can calculate the overall sample mean as the simple mean of the observations y_{jk},

$$\frac{1}{n} \sum y_{jk} = \overline{y}_{\text{strat prop}},$$

where $n = \sum n_j$. If, however, the strata differ in variability, an allocation that samples more heavily from more variable strata can reduce the standard error of the overall estimate, at the cost of requiring weighted calculations. An intuitive feel for this idea derives from considering two strata, one in which all the units are identical on the variable of interest and the other in which the units differ over a wide range. A single observation from the non-varying stratum is sufficient to estimate the stratum mean, whereas the more variable the units in the varying stratum, the more observations we need to pin down the mean. Cost considerations also affect allocation. Other things equal, it makes sense to draw fewer units from strata where observations are expensive than from strata where observations are cheaper. In practice the design of a sample balances variability against cost.

Stratified sampling aims to divide the population into strata that are as different from one another as possible and as internally homogeneous as possible; *cluster sampling* does the opposite. Here a sampler divides the population into subgroups that are typically compact geographically but as diverse as possible internally—ideally each cluster should mirror the entire population. Then the sampler uses a probability sampling method to choose several of the clusters to constitute the final sample. Cluster samples often reduce interviewer travel costs.

Most large-scale surveys combine clustering, stratification, and systematic sampling in a complex procedure known as *multistage area probability sampling*. One motivation for this procedure is the lack of any complete list of all individuals in the United States to serve as a frame for a national survey. Let us consider how multistage area probability sampling is accomplished for the CPS, for which the unit chosen and visited is a housing unit, but for which inferences are made about individuals.

The CPS designates as *primary sampling units* (PSUs) counties or groups of contiguous counties (1973 PSUs in 1980). The PSUs are stratified using such criteria as Metropolitan Statistical Area (MSA) or not, principal industries, and racial composition. The 333 PSUs with the largest population are designated as self-representing and included in the sample with certainty. For each remaining stratum, one PSU is selected with probability proportional to population as reported in the latest census, with an attempt to include at least one PSU from each state.

Each PSU consists of enumeration districts (EDs), constructed from the previous census. The EDs are sampled systematically with a random start, being assigned selection probabilities proportional to size. The selected EDs, if they have good address listings from the previous census, have their address lists improved, and address sampling is used to choose the ultimate sampling units (USUs), clusters expected to contain approximately four housing locations each. A further sample of building permits provides coverage of units constructed since the previous census.

If the ED does not have a good address listing, USUs are chosen in two further stages of area sampling. First, the ED is divided into blocks or chunks expected to contain two to five USUs. One of these is chosen and is visited so that a list can be made of all current living quarters. This list is subsampled to form the USU.

At each of these steps, sampling probabilities are calculated so that the sample remains self-weighting for the calculation of means and proportions. Because the number of people in a household varies, issues of weights become important in CPS when means or proportions are calculated using the individual as the unit of analysis.

4. SOME ISSUES IN SURVEYS BY TELEPHONE

Although the popular image of a survey taker is probably that of an earnest female interviewer ringing the doorbell of one of the chosen, in many surveys no interviewer goes to anyone's door. Some surveys are conducted by mail, and many are conducted by telephone. Some special problems of sampling arise in surveying by telephone. The most obvious is that not all families have a phone. Thornberry and Massey (1988) report that in 1985-86 some 7.2% of U.S. households had no phone; but the percentage of individuals living in households without a phone is much higher for certain special groups, such as blacks, young children and youths 17–24, those separated or divorced, and those with low incomes. This lack of telephone coverage introduces a selection bias. Perhaps the selection bias is not severe if the target population is the entire U.S. population. But if the target population is one of these special groups, using telephone interviewing gives many members of the target population zero probability of being selected into the sample and hence would create severe selection bias.

If telephone directories provide the frame for sampling, several additional problems immediately arise. Like any list, a directory is out of date as soon as it is published, because it contains numbers of people who have moved away and omits numbers of those who have moved into an area. But more important, many telephone subscribers request that the phone company not publish their phone numbers. The proportion of non-published numbers has been growing in recent years and by now constitutes over half the residential telephones in some urban areas. Hence, using a telephone directory as a frame also introduces selection bias. Survey researchers have developed a procedure known as *random digit dialing* (RDD) to overcome this bias—in effect, within an area code RDD generates a seven-digit random number and thus uses as a sampling frame all 10^7 possible listed and unlisted telephone numbers.

This conceptual description of RDD glosses over several difficulties. Most important, not all 10^7 telephone numbers represent working telephones, and many working numbers are nonresidential phones not useful for a household survey. Thus if numbers were generated by SRS from the 10^7 numbers, the majority of calls made would be useless. A refinement of the frame is usually made by obtaining from Bell Communications Research a list of working area codes and three-digit prefixes (or exchanges) and randomly appending the final four digits. Even with this refinement some 75%–80% of the generated numbers are not assigned to households (Lepkowski 1988), so again many dialings are unproductive.

To decrease the number of such unproductive dialings, survey researchers have developed clustering procedures in which they use "banks" of phone numbers within an area code and prefix, taking advantage of the telephone company's practice of assigning successive telephone numbers. A bank is defined by one or more digits of the suffix. Thus one can speak of banks of 1000, 100, or 10 numbers. A bank is chosen at random, and a phone number within that bank is generated and dialed. If the number represents an eligible household, then sampling continues from that bank; if the primary number is ineligible, no further numbers are generated from that bank. These ideas originated with Mitofsky (1970) and Waksberg (1978) and are well explicated by Lepkowski (1988).

5. NONSAMPLING ERRORS

So far in this chapter we have assumed an ideal world. Among other things, we assumed that the frame is an accurate representation of the population to which we want to generalize, that everyone chosen for the sample provides the data requested, that the researcher and respondent share the same definition of all concepts involved, that respondents remember correctly and tell the truth, and that nobody makes a mistake in copying down the answer. Violations of these assumptions— *nonsampling errors*—have attracted much interest, for in some ways nonsampling errors are harder to understand and control than sampling errors. Some nonsampling errors are essentially random—copying errors, for example—and they tend to cancel out as the sample size increases, though they will increase the variance of the estimates. Other nonsampling errors, such as memory

errors and systematic coding errors, tend to accumulate and cannot be decreased just by increasing the size of the sample. As James A. Davis (1975, p. 42) puts it, "$\sqrt{n}$ wrongs do not make a right."

Some nonsampling errors are essentially random, and they tend to cancel out as the sample size increases, though they will increase the variance of the estimates. Other nonsampling errors tend to accumulate and cannot be decreased just by increasing the size of the sample.

Nonsampling errors can be subdivided into *nonresponse errors* (people are left out of the frame, left out of the sample, or do not answer specific questions) and *response* or *measurement errors* (answers are obtained but are in some sense "wrong").

5.1 Response Errors

The investigation of response errors has a long history in survey research. Issues explored include

- the tendency of respondents to give socially desirable answers

- failures of memory

- differences in interviewer behavior across respondents or differing respondent reactions to interviewers by such variables as race, gender, and social class

- mismatch between the meaning of the question as intended by the researcher and the meaning understood by the respondent

- differences in answers associated with the mode of administering the interview (by mail, by phone, or in person)

- characteristics of the question itself (long or short, open-ended or with explicit response alternatives, presented in a positive or negative manner, and so on)

- the context of other questions in which a particular question is embedded and the ordering of questions in an interview or questionnaire

- whether the answer is given by the person directly concerned or by a proxy respondent.

Several excellent reviews of this literature include Dalenius (1977), Kahn and Cannell (1968), Mosteller (1968), Schuman and Presser (1981), Sudman and Bradburn (1974), and Turner and Martin (1984). In recent years research has begun in a movement that aims to bring the theories and methodologies of the cognitive sciences, especially cognitive psychology, to bear on the understanding and solution of problems of response errors. An early report on the movement is contained in Jabine, Straf, Tanur, and Tourangeau (1984); Jobe and Mingay (1991) provide a review of the literature.

5.2 Nonresponse Errors

We know that those who do not answer some or all questions in a survey or who are never at home to an interviewer are different from those who answer or are at home, at least in terms of refusing to answer or being away from home. It is likely that they are different in other ways as well. And if these ways involve the variable(s) that the survey is trying to measure (say, income or political opinion), then the self-selection bias discussed in Section 1 endangers the validity of the results of the survey.

It is useful to distinguish between unit nonresponse and item nonresponse. In *unit nonresponse*, entire sets of data are missing for potential respondents because the respondents were missed in the field (for example, they were never at home), were missed in the frame (for example, they did not have telephones), or refused to participate. Item nonresponse occurs when an individual's answers to some parts of a survey instrument are missing or are inconsistent (for example, if wage income plus interest income plus income from other sources is greater than total income) and so are edited out in the data-cleaning process and must be replaced by a more consistent set of answers.

Both item and unit nonresponse are high and getting higher, even in surveys under government sponsorship—a trend that has continued for the past three decades (Groves 1989). Refusal rates for the CPS rose from 1.8 percent in 1968 to 2.5 percent in 1976; for the National Health Interview Survey, from 1.2 percent to 2.1 percent in the same time period (Panel on Privacy and Confidentiality as Factors in Survey Response 1979). These numbers engender particular concern because (1) both

of these surveys are conducted by the U.S. government, (2) extensive and increasing efforts are mounted to reach respondents initially not found at home, and (3) each one percent of the American population represents over two million individuals. The problem is not confined to the United States, however. Results of the Swedish government Labor Force Survey show that refusals rose from 1.2 percent in 1970 to 3.9 percent in 1977 (Dalenius 1979).

Even the U.S. census, to which a response is required by law, is not immune. In the 1970 U.S. census, data had to be imputed (filled in) for such items as age (4.5 percent of the respondents) and total family income (20.7 percent of families, although many of these families reported several components of income) (Bailar and Bailar 1979). Closer to unit nonresponse, it is estimated that the 1970 census undercounted by 2.5 percent (or about five million people), even after adding people to the count whenever there was a shred of evidence to do so. (Housing and post office checks by the Census Bureau on a sample basis pointed to some occupied buildings in which no residents were counted. Thus the Bureau could add some five million people who had not filled in census forms before the estimate of the five million undercount was calculated.) The problem of undercounting or nonresponse in the 1980 and 1990 censuses has been a major source of legal challenges.

Nonresponse is an even greater problem in nongovernment surveys. In surveys with a variety of sponsors, dealing with a variety of populations, and using a variety of definitions of nonresponse, one study found nonresponse ranging from a low of about 5 percent to a high of about 87 percent (Panel on Privacy and Confidentiality as Factors in Survey Response 1979). Without substantial efforts to curb nonresponse, lowered response rates in major national data collection programs may render survey results practically and scientifically useless. Thus, the vigorous scientific activity being devoted to developing methods for reducing nonresponse, for adjusting for it when it does occur, and for properly analyzing the resulting data is crucial to continued good quality data from surveys.

Nonresponse in the sense of noncoverage in the frame can be unintentionally introduced in the design stage (Morris 1979). For example, a design based on imperfectly measured variables or those that are subject to random change will exclude some part of the population. A frame confined to low-income people, for example, will exclude those whose incomes in the critical year were "accidentally" higher than their permanent incomes. (Of course, those with "accidentally" lower incomes will be mistakenly included.) Similarly, a frame constructed to tap large concentrations of a target group will often miss atypical members of that group. Thus a frame using low-income census tracts to reach low-income people would miss low-income people living in high-income tracts. Of course, telephone surveys miss non-telephone households.

Certainly the preferred method of dealing with nonresponse is to keep it from happening. Thus survey researchers have developed a battery of techniques with the general aim of encouraging chosen respondents to participate or of systematically substituting other informants or respondents in the field. Callbacks, incentives, and enlisting the cooperation of local governments are among the techniques of encouragement. (But survey researchers recognize that sometimes extreme efforts to decrease nonresponse may degrade the quality of the data as reluctant respondents give less valid responses.) Permitting proxy response reduces nonresponse at the same time that it reduces callbacks.

Without substantial efforts to curb nonresponse, lowered response rates in major national data collection programs may render survey results practically and scientifically useless.

Despite the best efforts of survey designers and field staff, nonresponse, both unit and item, frequently occurs and must be taken into account. What, then, can be done after the fact, to adjust for appreciable nonresponse? It is logically impossible to do nothing. Simply to drop the nonresponding units from the sample is to do something, for then estimation procedures tacitly assume that nonresponders are just like responders and that the results of the survey would not have changed had they responded. Doing nothing implies a very specific but simple model: the forces that prevented some people from responding are unrelated to the variables of interest, so that the distribution of nonrespondents on these variables is no different from the distribution of respondents. Similarly, more complex techniques for dealing with missing data also require implicit or explicit models of the causes of nonresponse and hence of the distribution of nonrespondents. The usual assumption is that nonrespondents are distributed like some subset of the

respondents with measured characteristics (covariates) similar to those of the nonrespondents, but sometimes the assumption is that nonrespondents differ from respondents in systematic ways (as would be true if, for example, the probability that people would report their income were inversely proportional to income).

Many techniques for dealing with missing data have been developed. See, for example, Bailar (1978), Morris (1979), Brewer and Sarndal (1979), Little and Rubin (1979), Kalsbeek (1980) and especially Madow, Nisselson and Olkin (1983). Some techniques reweight aggregations of data to take into account missing observations, and others "fill in the blanks," creating pseudo-observations in place of the missing ones. In either case, the analyst must recognize that the data have been adjusted for nonresponse and that such adjustments affect estimates of the accuracy of quantities derived from the data.

A common means of weighting for missing data is *ratio estimation*. It uses information derived from other studies to improve estimation. Assume that the quantity we wish to estimate is $\overline{Y}$ (for example, the average income for the population) and that it will be estimated by the sample mean $\overline{y}$ (the average income for those in the sample). Assume also that we know that Y is related to another variable, X (say, the number of people per room in living quarters), for which we know both the mean for the respondents in the sample, $\overline{x}$ (mean number of people per room in the sample) and the mean for the total population, $\overline{X}$, from another source, such as the census. If we make the additional assumption that the ratio of the mean number of people per room in the sample to the mean number of people per room in the population is the same as the corresponding ratio of mean income in the sample to mean income in the population $(\overline{x}/\overline{X} = \overline{y}/\overline{Y})$, then we can use this relation to adjust $\overline{y}$ to $\overline{y}^* = \overline{y}(\overline{X}/\overline{x})$. Deming (1968) and Cochran (1977) present properties of this estimator and several related ones. Even if the assumptions are only approximately correct, ratio estimation usually improves accuracy; as a bonus, ratio estimation usually reduces sampling error.

Tupek and Richardson (1978), for example, used a ratio adjustment for nonresponse to correct for nonresponse bias in the 1975 Survey of Scientific and Technical Personnel. They found that large firms were least likely to respond to the survey. They knew the total number of employees in the firms in each size stratum from other sources, and the ratio of scientific and technical employees to total employees remained constant. Hence, Tupek and Richardson were able to use the ratio of total employees in the reporting firms in the stratum to total employees in all firms in the stratum to adjust the estimated number of scientific and technical personnel in each stratum.

Techniques that fill in missing values individually for item nonresponse are called *imputation techniques*. Such techniques assume that the value of the missing item can be estimated from values of other items for that respondent. One such technique uses the other items as variables in a regression function, either derived from the data at hand or available from outside sources. Such a procedure must assume (or fit) a particular functional form for the relation of the missing item to the other variables (covariates).

In the days before high-speed computers, survey analysts often filled in blanks caused by item nonresponse from tables put together from outside sources. Such a table might specify that, if the respondent was a married white female between the ages of 30 and 45 who did not answer how many children she had, she should be "assigned" two children. This so-called "cold-deck" procedure, of course, assigned the same number of children to all missing values for women in a specific marital status-race-age group. With the advent of high-speed computers, more flexible procedures became possible.

These *hot-deck* procedures fill in the missing value for the item from the value appearing for another respondent in the same survey who is "similar" to the respondent with missing data. "Similar" is defined by the variables thought to influence the one missing (for example, for number of children, these variables might still be marital status, race, and age), and all respondents who are the same on these variables are said to constitute an "adjustment class" (Sande 1979). Hot-deck procedures make no assumptions about the functional form by which the variables defining the adjustment class determine the missing item, only that they do. Among the tremendous variety of these hot-deck procedures, the simplest uses the value of the item that occurred in the previous unit processed in that adjustment class. Other variations (made possible by advances in computer science, random access, and dynamic creation of the adjustment classes) choose a donor within the adjustment class on criteria of nearness on further important variables or introduce randomness into the process of choice of a donor (Sande 1979).

Analysts must exercise care when making estimates from data that have been partially imputed, because the imputation changes the estimated accuracy of the estimates. Further, the sample size for any item is the number of respondents actually giving data for that item, and so imputation does not increase sample size.

A rather new idea is a process of multiple imputation (Rubin 1987). Here the analyst repeatedly uses an imputation method to fill in missing data. Each time the complete data set is imputed, an estimate is made of the quantity of interest. One can then examine the distribution of these estimates to see whether, or how much, they vary with different imputed data sets. If several different assumptions about the causes of nonresponse are plausible, a set of multiple imputations might be carried out using each assumption as the model to determine the imputation method. The differences among these sets of estimates allow for exploration of the sensitivity of the estimation to the model assumed for nonresponse. The justification and interpretation of this multiple imputation procedure have their roots in Bayesian statistical theory.

6. CONCLUSION

This chapter has attempted to build a bridge between the ideas usually taught in statistics courses and some aspects of everyday sampling and survey practice. Most introductory statistics courses, because of time pressures and the potential confusion arising from the consideration of more complicated cases, concentrate on simple random sampling and assume that the population size is much larger than the sample size. Although this simplification does make for easier exposition, teachers and students should understand that most survey data arise from sample designs more complicated than simple random sampling and that finite populations are common in practice. Inference from a sample is based on a probability model, exemplified by the discussion of the sampling distribution of the sample mean in Section 3. But as the sample design changes, so do the appropriate probability model and consequently the proper inference methods. The specific recipes for inference learned in introductory courses are thus usually appropriate only in simple cases. The reasoning behind those recipes, however, transfers to the more complicated cases encountered in practice.

Another aim of this chapter is to stress the importance of the nonmathematical aspects of surveys and, by extension, of practical statistics more generally. Obtaining complete frames, choosing strata, wording questions, reducing nonresponse, and many other steps in fielding a survey are essential practical skills and entail a good deal of informed judgment. Confidence intervals and other measures of "margin of error" reported in scientific work and in the media account only for the uncertainty introduced by sampling error, not for any further uncertainty stemming from nonresponse, from wording choices, and for all the other possible sources of nonsampling error. Newspapers such as the *New York Times* that report survey results have become sensitive to issues of nonsampling errors and now often couple their citation of a margin of error with a warning that the practical exigencies of conducting a survey may introduce further errors.

The bibliography below offers opportunities for further reading. Works on applications of sampling and on nonsampling errors have been cited in text in a manner that should give clear guidance to their contents. The bibliography also lists several classic treatments of sampling. Perhaps the most accessible are Slonim (1960) and Williams (1978). Cochran (1977) and the second volume of Hansen, Hurwitz, and Madow (1953) present a full theoretical development. More applied are Deming (1960), Hansen, Hurwutz, and Madow (1953, Volume 1), Kish (1965), and Yates (1960).

Acknowledgments

Portions of this chapter are based on earlier publications to which the author has contributed, especially Tanur (1984) and Fienberg and Tanur (1983). My thanks for editorial assistance to Stephen Fienberg, David Moore, and especially David Hoaglin.

Judith M. Tanur is Professor of Sociology at the State University of New York at Stony Brook. She received a B.S. in psychology and an M.A. in mathematical statistics, both from Columbia University, and a Ph.D. in sociology from the State University of New York at Stony Brook. Her main interests are in cognitive aspects of surveys, parallels between sample surveys and designed experiments, and the application and teaching of statistics in the social sciences. She is a Fellow of the American Association for the Advancement of Science and of the American Statistical Association and an elected member of the International Statistical Institute. She is a recipient of the President's and Chancellor's awards for excellence in teaching at Stony Brook.

REFERENCES

Bailar, B. A. and Bailar, J. C., III (1979), "Comparison of the Biases of the Hot Deck Procedure with an Equal-Weights Imputation Procedure," in Panel on Incomplete Data (1979), 422–447.

Bailar, J. C., III (1978), "Discussion," in *Imputation and Editing of Faulty or Missing Survey Data*, eds. F. Aziz and F. Scheuren, Washington, DC: U.S. Department of Commerce, Bureau of the Census.

Brewer, K. R. and Särndal, C. E. (1979), "Six Approaches to Enumerative Survey Sampling," in Panel on Incomplete Data (1979), 363–368.

Bryson, M. C. (1976), "The *Literary Digest* Poll: Making of a Statistical Myth," *The American Statistician*, **30**, 184–185.

Chapman, D. G. (1989), "The Plight of the Whales," in *Statistics: A Guide to the Unknown*, third edition, eds. J. M. Tanur, F. Mosteller, W. H. Kruskal, E. L. Lehmann, R. F. Link, R. S. Pieters, and G. R. Rising, Pacific Grove, CA: Wadsworth and Brooks/Cole, 60–67.

Childers, D., Diffendal, G., Hogan, H., Schenker, N., and Wolter, K. (1987), "The Technical Feasibility of Correcting the 1990 Census," *1987 Proceedings of the Social Statistics Section*, 36–45, Alexandria, VA: American Statistical Association.

Cochran, W. G. (1977), *Sampling Techniques*, third edition, New York: John Wiley.

Dalenius, T. (1977), "Bibliography on Nonsampling Errors in Surveys," *International Statistical Review*, **45**, 71–89, 181–197, 303–317.

—— (1979), " Informed Consent or R.S.V.P.," in Panel on Incomplete Data (1979), 94–134.

Davis, J. A. (1975), "Are Surveys Any Good, and If So, for What?" in *Perspectives on Attitude Assessment Surveys and Their Alternatives*, eds. H. W.

Sinaiko and L. A. Broedling, Washington, DC: Manpower Research and Advisory Services, Smithsonian Institution—Proceedings of a Conference Held at The Bishop's Lodge, Santa Fe, New Mexico, April 22–24, 1975.

Davis, J. A. and Smith, T. W. (1988), *General Social Surveys 1979-1988: Cumulative Codebook*, Chicago: National Opinion Research Center.

Deming, W. E. (1960), *Sample Design in Business Research*, New York: John Wiley.

—— (1968), "Sample Surveys I: The Field," in *International Encyclopedia of the Social Sciences*, ed. D. L. Sills, New York: Macmillan and The Free Press. Reprinted and updated in 1978 in *International Encyclopedia of Statistics*, eds. W. H. Kruskal and J. M. Tanur, New York: Macmillan and The Free Press.

—— (1986), *Out of the Crisis*, Cambridge, MA: MIT Center for Advanced Engineering Study.

Ferber, R., Sheatsley, P., Turner, A., and Waksberg, J. (1980), *What Is a Survey?* Washington, DC: American Statistical Association.

Fienberg, S. (1971), "Randomization and Social Affairs: The 1970 Draft Lottery," *Science*, **171**, 255–261.

Fienberg, S. and Tanur, J. (1983), "Large Scale Social Surveys: Perspectives, Problems, and Prospects," *Behavioral Science*, **28**, 135–153.

Fisher, R. A. (1930), "Inverse Probability," *Proceedings of the Cambridge Philosophical Society*, **26**, 528–535.

—— (1932), "Inverse Probability and the Use of Likelihood," *Proceedings of the Cambridge Philosophical Society*, **28**, 257–261.

—— (1933), "The Concepts of Inverse Probability and Fiducial Probability Referring to Unknown Parameters," *Proceedings of the Royal Society of London, Series A*, **139**, 343–348.

Gini, C. (1928), "Une application de la méthode répresentative aux materiaux du dernier recensement de la population Italienne," *Bulletin of the International Statistical Institute*, **23** (Liv. 2), 198–215.

Gini, C. and Galvani, L. (1929), "Di una applicazione del metodo rappresentive all'ultimo censimento Italiano della popolazione (10 decembri, 1921)," *Annali di Statistica*, Series 6, **4**, 1–107.

Groves, R. M., Biemer, P. P., Lyberg, L. E., Massey, J. T., Nicholls, W. L., II, and Waksberg, J. (eds.) (1988), *Telephone Survey Methodology*, New York: John Wiley.

Groves, R. M. (1989), *Survey Costs and Survey Errors*, New York: John Wiley.

Hamilton, D. P. (1990), "Census Adjustment Battle Heats Up," *Science*, **248**, 807–808.

Hansen, M. H., Hurwitz, W. N., and Madow, W. G. (1953), *Sample Survey Methods and Theory* (2 vols.), New York: John Wiley.

Hite, Shere (1987), *Women and Love: A Cultural Revolution in Progress*, New York: Alfred A. Knopf.

Hogan, H. and Wolter, K. (1988), "Measuring Accuracy in a Post-enumeration Survey," *Survey Methodology*, **14**, 99–116.

Jabine, T., Straf, M., Tanur, J., and Tourangeau, R. (eds.) (1984), *Cognitive Aspects of Surveys: Building a Bridge Between Disciplines*, Washington, DC: National Academy Press.

Jobe, J. and Mingay, D. (1991), "Cognition and Survey Measurement: History and Overview," *Applied Cognitive Psychology*, **5**, 175–192.

Kahn, R. L. and Cannell, C. F. (1968), "Interviewing in Social Research," in *International Encyclopedia of the Social Sciences*, ed. D. L. Sills, New York: Macmillan and The Free Press. Reprinted and updated in 1978 in *International Encyclopedia of Statistics*, eds. W. H. Kruskal and J. M. Tanur, New York: Macmillan and The Free Press.

Kalsbeek, W. D. (1980), "A Conceptual Review of Survey Error Due to Nonresponse," *1980 Proceedings of the Section on Survey Research Methods*, 131–136, Washington, DC: American Statistical Association.

Keyfitz, N. (1957), "Estimates of Sampling Variance Where Two Units Are Selected from Each Stratum," *Journal of the American Statistical Association*, **52**, 503–510.

Kish, L. (1965), *Survey Sampling*, New York: John Wiley.

Kish, L. and Frankel, M. R. (1974), "Inference from Complex Samples" (with discussion), *Journal of the Royal Statistical Society, Series B*, **36**, 1–37.

Kramer, F. D. (ed.) (1988), *From Quality Control to Quality Improvement in AFDC and Medicaid*, Washington, DC: National Academy Press.

Leon, C. B. and Rones, P. L. (1989), "How the Nation's Employment and Unemployment Estimates Are Made," in *Statistics: A Guide to the Unknown*, third edition, eds. J. M. Tanur, F. Mosteller, W. H. Kruskal, E. L. Lehmann, R. F. Link, R. S. Pieters, and G. R. Rising, Pacific Grove, CA: Wadsworth and Brooks/Cole, 218–226.

Lepkowski, J. M. (1988), "Telephone Sampling Methods in the United States," Chapter 5 in *Telephone Survey Methodology*, eds. R. M. Groves, P. P. Biemer, L. E. Lyberg, J. T. Massey, W. L. Nicholls II, and J. Waksberg, New York: John Wiley.

Link, R. F. (1989), "Election Night on Television," in *Statistics: A Guide to the Unknown*, third edition, eds. J. M. Tanur, F. Mosteller, W. H. Kruskal, E. L. Lehmann, R. F. Link, R. S. Pieters, and G. R. Rising, Pacific Grove, CA: Wadsworth and Brooks/Cole, 104–112.

Little, R. J. A. and Rubin, D. B. (1979), "Six Approaches to Enumerative Survey Sampling, Discussion," in Panel on Incomplete Data (1979), 515–520.

Literary Digest (1936), issue of August 22, p. 3.

Madow, W. G., Nisselson, H., and Olkin, I. (1983), *Incomplete Data in Sample Surveys* (3 volumes), New York: Academic Press.

Mahalanobis, P. C. (1944), "On Large-Scale Sample Surveys," *Philosophical Transactions of the Royal Society*, **231(B)**, 329–451.

McCarthy, P. J. (1966), *Replication, an Approach to the Analysis of Data from Complex Surveys*, Series 2, No. 14, U.S. Department of Health, Education and Welfare, National Center for Health Statistics, Washington, DC: U.S. Government Printing Office.

—— (1969), *Pseudoreplication, Further Evaluation and Application of the Balanced Half-Sample Technique*, Series 2, No. 31, U.S. Department of Health, Education and Welfare, National Center for Health Statistics, Washington, DC: U.S. Government Printing Office.

Mitofsky, W. (1970), *Sampling of Telephone Households*, Unpublished memorandum, CBS.

Morris, C. (1979), "Nonresponse Issues in Public Policy Experiments, with Emphasis on the Health Insurance Study," in Panel on Incomplete Data (1979), 448–470.

Mosteller, F. (1968), "Errors I: Nonsampling Errors," in *International Encyclopedia of the Social Sciences*, ed. D. L. Sills, New York: Macmillan and The Free Press. Reprinted and updated in 1978 in *International Encyclopedia of Statistics*, eds. W. H. Kruskal and J. M. Tanur, New York: Macmillan and The Free Press.

Neter, J. (1989), "How Accountants Save Money with Sampling," in *Statistics: A Guide to the Unknown*, third edition, eds. J. M. Tanur, F. Mosteller, W. H. Kruskal, E. L. Lehmann, R. F. Link, R. S. Pieters, and G. R. Rising, Pacific Grove, CA: Wadsworth and Brooks/Cole, 151–160.

Neyman, J. (1934), "On the two different aspects of the representative method: The method of stratified sampling and the method of purposive selection," *Journal of the Royal Statistical Society (A)*, **109**, 558–606.

Panel on Incomplete Data of the Committee on National Statistics, National Research Council, (1979) *Symposium on Incomplete Data: Preliminary Proceedings*, Washington, DC: U.S. Department of Health, Education and Welfare, Social Security Administration, Office of Policy, Office of Research and Statistics.

Panel on Privacy and Confidentiality as Factors in Survey Response (1979), *Privacy and Confidentiality as Factors in Survey Response*, Washington, DC: National Academy of Sciences.

Press, S. J. (1989), "Statistics in Jury Selection: How to Avoid Unfavorable Jurors," in *Statistics: A Guide to the Unknown*, third edition, eds. J. M. Tanur, F. Mosteller, W. H. Kruskal, E. L. Lehmann, R. F. Link, R. S. Pieters, and G. R. Rising, Pacific Grove, CA: Wadsworth and Brooks/Cole, 87–92.

The RAND Corporation (1955), *A Million Random Digits with 100,000 Normal Deviates*, Glencoe, IL: Free Press of Glencoe.

Rossi, P. H., Wright, J. D., and Anderson, A. B. (eds.) (1983), *Handbook of Survey Research*, Orlando, FL: Academic Press.

Rubin, D. B. (1987), *Multiple Imputation for Nonresponse in Surveys*, New York: John Wiley.

Sande, G. (1979), "Hot Deck Discussion—Replacement for a Ten Minute Gap," in Panel on Incomplete Data (1979), 481–483.

Schuman, H. and Presser, S. (1981), *Questions and Answers in Attitude Surveys: Experiments on Question Form, Wording, and Context*, New York: Academic Press.

Slonim, M. J. (1960), *Sampling*, New York: Simon and Schuster. (Originally published as *Sampling in a Nutshell*.)

Stephenson, C.B. (1978), *A Comparison of Full-Probability and Probability-with-Quotas Sampling Techniques in the General Social Survey*, General Social Survey Technical Report No. 5, Chicago: National Opinion Research Center.

Stuart, A. (1968), "Sample Surveys II: Nonprobability Sampling," in *International Encyclopedia of the Social Sciences*, ed. D. L. Sills, New York: Macmillan and The Free Press. Reprinted and updated in 1978 in *International Encyclopedia of Statistics*, eds. W. H. Kruskal and J. M. Tanur, New York: Macmillan and The Free Press.

Sudman, S. (1967), *Reducing the Cost of Surveys*, Hawthorne, NY: Aldine.

Sudman, S. (1983), "Applied Sampling," Chapter 5 in *Handbook of Survey Research*, eds. P. H. Rossi, J. D. Wright, and A. B. Anderson, Orlando, FL: Academic Press.

Sudman, S. and Bradburn, N. M. (1974), *Response Effects in Surveys: A Review and Synthesis*, Hawthorne, NY: Aldine.

Tanur, J. M. (1984), "Methods for Large-Scale Surveys and Experiments," Chapter 1 in *Sociological Methodology 1983–1984*, ed. Samuel Leinhardt, San Francisco: Jossey-Bass.

Thornberry, O. T., Jr., and Massey, J. T. (1988), "Trends in United States Telephone Coverage Across Time and Subgroups," Chapter 3 in *Telephone Survey Methodology*, eds. R. M. Groves, P. P.

Biemer, L. E. Lyberg, J. T. Massey, W. L. Nicholls II, and J. Waksberg, New York: John Wiley.

Tupek, A.R. and Richardson, W. J. (1978), "Use of Ratio Estimates to Compensate for Nonresponse Bias in Certain Economic Surveys," in *Imputation and Editing of Faulty or Missing Survey Data*, eds. F. Aziz and F. Scheuren, Washington, DC: U.S. Department of Commerce, Bureau of the Census.

Turner, C. and Martin, E. (eds.) (1984), *Surveying Subjective Phenomena*, New York: Russell Sage Foundation.

U.S. Bureau of Labor Statistics (1988), *BLS Handbook of Methods*, Bulletin 2285, Washington, DC: U.S. Government Printing Office.

Waksberg, J. (1978), "Sampling Methods for Random Digit Dialing," *Journal of the American Statistical Association*, **73**, 40–46.

Williams, B. (1978), *A Sampler on Sampling*, New York: John Wiley.

Yates, F. (1960), *Sampling Methods for Censuses and Surveys*, third edition, London: Charles Griffin.

Chapter 5

The Statistical Approach to Design of Experiments

Ronald D. Snee
E. I. du Pont de Nemours & Company

Lynne B. Hare
Thomas J. Lipton Company

1. ACQUIRING KNOWLEDGE

Statistical design of experiments is an essential part of the scientific method of inquiry, and it is an extremely powerful approach to experimentation. Using the scientific method, we begin with conjectures based on prior knowledge and spurred by dissatisfaction with current understanding. This is the inductive step. We design a data collection plan; that is, we *design the experiment*. Then we collect the data, we analyze the data to separate real effects from chance fluctuations, and we draw conclusions. These are the deductive steps.

The power of experimental design technology stems from the efficiencies that it affords. This framework allows us to measure the simultaneous effects of many factors (or variables) as they act independently or in concert. Thus, one well-planned experiment, incorporating many factors, can give much more information than many, many small experiments that study these same factors individually. Examples in subsequent sections show how this is done.

The first applications of experimental design led to outstanding advances in biology and agriculture. The last few decades have witnessed diverse applications to many fields of scientific endeavor, including pharmacology, behavioral science, education, medicine, and sociology.

Industrial applications took place as early as the 1950s, but the 1980s brought a real industrial awakening to the power of planned experimentation. In the West, this awakening was caused by painful losses in global markets. The United States, for example, lost its competitive edge in industries such as steel, automobiles, electronics, and textiles to nations in the Pacific basin. It is ironic that, although most of the techniques of experimental design were developed in the West, its practice is most often followed in the East.

The statistical design of experiments guides us in deciding what data and how much data should be collected to understand variability and help identify ways to control it.

Ignoring these powerful tools is part of the reason for the losses, for only through the use of experimental design techniques can we hope to get effective answers to questions such as

- How can we improve process yield?

- How can we reduce scrap and rework?

- How can we evaluate the efficacy and safety of a new drug?

- How can we design a new product that will not cause harm to the environment?

- How can we design a new medical device that will give accurate measurements when used by patients at home?

- How can we design a food recipe that will perform well regardless of factors beyond our control, such as varying water hardness and oven temperatures?

Much of Western industry has recognized that they must work in new ways, using new approaches to problem-solving and learning, if they are to regain a competitive edge. Fundamentally, customer satisfaction has become a requirement for survival. Because customers insist on it, industry is being forced to direct it's attention to quality as an overriding concern. Notice that a concern for quality is at the heart of each of the questions above. The key question is how industry can provide products and services that will better satisfy the needs of customers so that, eventually, they can regain the competitive edge. Well-known authors and sources have identified statistics as a body of knowledge that can help provide answers (Deming 1986, Business Week 1987, Penzias 1989).

Experimentation seeks to determine how changes in the predictor variables or factors affect the responses.

How can statistics help? Statistics aids decision-making in the presence of variability. "Statistical thinking" recognizes that

- all work is a process,

- all processes contain variability, and

- it is almost always to our advantage to work to minimize that variability.

Minimizing variability comes from understanding the variability and working to identify ways to control it. The "statistical thinking" process is shown schematically in Figure 1 (Snee 1990). Notice that we use a generic definition of "process" to describe any series of events which can range from cell multiplication, to food manufacturing, to new drug development, to the way people work together.

In order to understand variability effectively, we must collect and analyze data. The "statistical design of experiments" guides us in deciding what data and how

much data should be collected to understand variability and help identify ways to control it.

2. STATISTICS HELP DECIDE WHAT EXPERIMENTS TO RUN

In general terms we think of a process as involving "input variables" and "control variables" that affect the "output." The input and control variables are typically referred to as "predictor variables" or "factors." The measurements made on the process outputs are called the "responses."

Experimentation seeks to determine how changes in the predictor variables or factors affect the responses. Statistical concepts and tools can help us decide what experiments to run so that we can determine which variables are important and to what extent they affect the response.

It is important to understand how the statistical approach to experimentation compares with other approaches. The "one-factor-at-a-time" (OFAAT) approach is the one practiced most widely by scientists and engineers. It is the approach that we all learned in our science classes. The procedure is: hold all factors constant except one, vary this factor over a range of values, and record how the responses of interest change. Hold this factor at some fixed level, pick out a different factor to vary, and observe the changes in the response while holding all other factors constant. This process continues until all the factors of interest have been studied or the desired results have been obtained.

The OFAAT approach is appealing because of its simplicity, but as R.A. Fisher (1935) pointed out many years ago, factorial experimentation has three advantages over the OFAAT approach. These are "greater efficiency," "greater comprehensiveness," and "a wider inductive basis for our conclusions." Further, the OFAAT approach makes two critical assumptions that are unrealistic. These assumptions may cause ineffective and inefficient experimentation.

The first assumption is that the relationship between the response and the predictor variables is very complex and that developing a predictive equation would require many experiments to ensure that no important information is overlooked. Although this assumption may be true over the full range of a predictor variable, practical considerations in most experimental situations dictate that predictor variables vary over a small portion

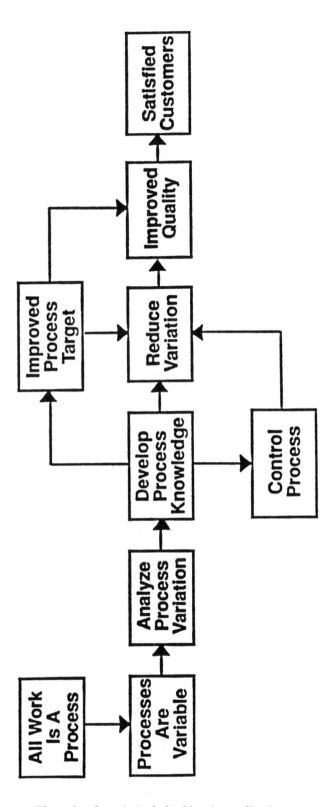

Figure 1: The role of statistical thinking in quality improvement.

| | Storage Period | |
Conditioning Time	SP_1	SP_2
	28	49
T_1	26	37
	30	38
	31	37
T_2	35	37
	31	29

Table 1: Easter lily experiment data: height (inches) at first bloom.

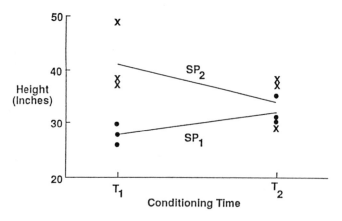

Figure 2: Height of Easter lilies.

of their possible range. The relationship between the response variable and the predictor variables, when the predictor variables are examined over narrow ranges, is usually approximated well by linear or quadratic functions. Use of such functions requires fewer experiments.

The second assumption of the OFAAT approach is that the effects of the factors are additive. This would be a simple world indeed, if that were always so. In many experimental situations, the combined effect of predictor variables does not equal the sum of the individual effects. The perception of strength of chocolate flavor, for example, depends on both the amount of chocolate present and the temperature used for baking the cake. For a low amount of chocolate, increasing the temperature increases the perception of chocolate flavor strength gradually. But for a high amount of chocolate, increasing temperature increases this perception at a much greater rate. The two predictor variables, amount of chocolate and baking temperature, act jointly and nonadditively to determine the perception of chocolate flavor strength. We call such joint action "interaction."

To discover the nature of interactions and to capitalize on their existence, we design experiments that vary the levels of the predictor variables jointly and systematically. Interaction is quite common in all kinds of processes, including those found in industry, as subsequent examples illustrate. To ignore the possibility of the existence of interactions, as the OFAAT approach does, is to overlook an opportunity for improving the process. This is the danger of the second assumption of the OFAAT approach.

Consider the following example (Ott 1975, pp. 229–232).

A horticulturist wanted to know whether the length of storage and the conditioning time after storage affected the height of Easter lilies at first bloom. He chose two storage periods (SP_1 and SP_2) and two conditioning times after storage (T_1 and T_2). The heights of three plants were measured at each of the four combinations of storage period and conditioning time in a 2×2 factorial design (Table 1). It was felt that this amount of replication would give an adequate estimate of the plant height at each of the experimental conditions. All other variables were held constant; however, the 12 plants were randomly distributed around the greenhouse to ensure that any unanticipated changes would not bias the results of the experiment.

One of the first steps in the analysis of any set of data is a graph of the data. Because the number of observations is small in this experiment, we can plot the individual data points, as shown in Figure 2. This graph helps illustrate the concept of interaction discussed above. The effect of conditioning time depends on storage period and vice versa. Conditioning time has a positive effect on plant height for the shorter storage period (SP_1) but a negative effect for the longer storage period (SP_2). If there were no interaction, increasing conditioning time would have the same effect for both storage periods. Another, equally valid way to describe this interaction is to note that storage period has a positive effect on Easter lily height when the conditioning time is short (T_1) but has little measurable effect at the longer conditioning time (T_2).

A second competitive approach to experimental design is the "Edisonian approach": Try everything, as Edison did. This approach is ruled out because no one has the time, money, or patience to do it!

We conclude that in the wide variety of experimental situations encountered in practice, ranging from agricultural field trials, to formulation of new drugs, to study of chemical processes, the responses are well-behaved in the experimental region of interest, the effects of factors being studied interact, and we cannot possibly "evaluate everything." The statistical approach to experimentation has been developed to deal with these real-life situations and constraints.

At the beginning of this chapter we noted the important role that statistics plays in the *process of experimentation*. Figure 3 depicts the "Iterative Nature of Experimentation," a process that develops knowledge by iterating between theories (ideas, conjectures, hypotheses) and facts (data collected in experiments).

The process begins with an idea, a conjecture, or a hypothesis to be tested. Data are subsequently collected by means of an experiment. These data are then compared to the original idea. The output of this comparison is a second or revised idea, hypothesis, or conjecture which, in turn, stimulates a new experiment that generates data to be compared to the revised idea. A further revised, refined idea develops, and the process continues until the desired knowledge or results are obtained.

The statistical approach plays a uniquely important role in the study of product quality.

Statistics plays two very critical roles in the process of experimentation. First, statistics helps us design the experiment so that the resulting data will enable us to evaluate the idea, conjecture, or hypothesis being investigated. Second, the statistical approach anticipates the analysis of the resulting data. This is one of the beauties of experimental design. As soon as we design the experiment, we know how we will analyze the data—assuming, of course, that the experiment is conducted as designed.

And so the process goes: idea 1, design 1, experiment 1, analysis 1, idea 2, design 2, experiment 2, analysis 2, idea 3, etc. Along the way we generate knowledge and understanding about the process being studied.

The statistical approach plays a uniquely important role in the study of product quality. Statistically designed experiments can help us identify which variables are critical to the control of our manufacturing processes. This methodology can also tell us how to plan experiments to convert customer suggestions into improved products and how to improve the performance of the processes we use to manufacture our products.

Statistical design of experiments is most valuable at the very beginning, when we design our products and manufacturing processes. The methodology enables us to build quality into our products and processes and avoid the waste associated with waiting until our customers, if they are kind, tell us of our faults. Design methodology enables us to create robust, rugged processes and products that are not affected by environmental variation and types of customer use (and abuse).

3. BENEFITS OF STATISTICALLY DESIGNED EXPERIMENTS

The statistical approach to design of experiments has broad applicability. We know of no field of study where it cannot be used. As illustrated by the discussion of product quality in the previous paragraph, the methodology is useful in almost any kind of investigation within any given field of study.

The benefits of the statistical approach to the process of experimentation are summarized in Table 2. Perhaps most important, the statistical approach provides a *system* for the design, analysis, and interpretation of results which produces *quality* data—data that satisfy the goals of the experiment. The approach enables the scientist or engineer to study the effects of a large number of variables that are of interest while controlling the influence of "nuisance" variables, which are not of interest but still influence the results.

In the analysis plan the statistical approach aims to extract the maximum possible information from the data, and it yields quantitative estimates for both the factor effects and their associated uncertainties. All phases of the analysis take into account the effects of experimental variation in the data and estimate them quantitatively.

In the critical interpretation phase, the statistical approach enables the user to develop trade-offs among multiple response variables. It is rare that a product or a process can be characterized by a single response variable. Textiles, for example, have color, texture, resilience, and several other properties. A single set of manufacturing settings is not likely to produce optimal values for all of the desired characteristics. Soft texture and resiliency in textiles, for example, seem opposed. When models are fit to data from designed experiments, we can predict the responses from the operating parameters, and we can use these predictive models to develop trade-offs representing compromises among responses. For example, we could produce the softest possible fabric with a specified resiliency.

The statistical approach also helps us identify the range

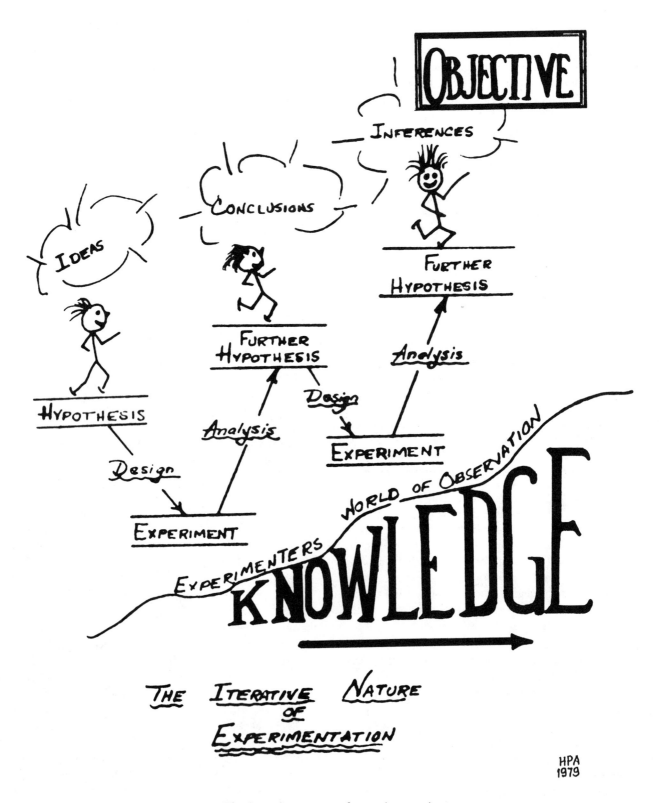

Figure 3: The iterative nature of experimentation.

- Systematic approach to experiment design and to analysis and interpretation of results

- Development of data that meet the needs of the study

- Ability to investigate the effects of a large number of variables

- Ability to control the influence of nuisance variables

- Efficient use of data

- Quantitative estimates of the effects of variables

- Quantitative estimate of experimental variation

- Ability to develop trade-offs among multiple response variables

- Identification of the range of validity of results

Table 2: Benefits of statistically designed experiments.

of validity of the results of an experiment. We are thus able to predict what will happen at different combinations of the factors, even when data were not collected at these different combinations.

Statistical methodology, therefore, provides an overall systematic approach to experimentation that helps set up experiments and plan the data collection process, offers analysis procedures to make sense of the data, and promotes valid interpretation of the results and sound decisions as a basis for action.

4. DESIGNING AN EXPERIMENT

The planning of an experiment must address some key issues. First we must have a statement of the objective which is agreed to by all concerned parties. This is often the most difficult part of the experiment.

Next we must consider the factors or variables to be studied and the responses or characteristics to be measured. We must determine whether the factors are quantitative (e.g., temperature, pressure) or qualitative (e.g., catalyst type, variety) and what levels of the factors we want to study. Quantitative factors are usually easier to study. Often, however, designs have only qualitative factors or a mixture of qualitative and quantitative factors. The two-level factorial designs to be discussed later are particularly effective in handling both qualitative and quantitative factors.

It is important that we have a good measurement of the responses or characteristics we are studying. Many experiments have failed because the response measurement was poor. Typically, a highly variable response re-

quires larger numbers of experimental runs at each combination of factors. It is not uncommon to work with attribute data such as "go-no-go" responses, percent-defective data, and subjective rating scales. Useful data can be obtained in these situations if the measurements are planned properly.

Next we must take into account the amount of resources (money, personnel, equipment) available for the study and the deadline for results. A good strategy avoids using all of the time and other resources on the first experiment. Reserve some resources to confirm the recommended changes developed from the initial experiments. The overall goal is to produce the needed results, information, etc. within the allotted time and money. The scope of the experimental program should be designed accordingly.

Two critical issues concerning the conduct of the experiment are *replication* and *randomization*. Replication refers to the number of times we will repeat each of the runs in the experimental design. Replication increases the precision of the estimates of the factor effects and provides an estimate of the experimental error that is used to calculate the uncertainty in the estimated factor effects.

Randomization is important to ensure that any unplanned changes that take place during the conduct of the experiment do not bias the results. Randomization is also critical to some of the assumptions of the statistical analysis. Complete randomization, in which the experimental runs are made in a completely random order, is often the preferred approach, but there are exceptions.

- Provides unbiased estimates of the factor effects and associated uncertainties

- Enables the experimenter to detect important differences

- Includes the plan for analysis and reporting of the results

- Gives results that are easy to interpret

- Permits conclusions that have wide validity

- Shows the direction of better results

- Is as simple as possible

Table 3: Characteristics of a good experimental design.

In agricultural research, for example, treatments such as applications of herbicides, pesticides, and fertilizers are allocated randomly within blocks or field plots. Several blocks are studied. This is done to minimize the impact of "noise" variables such as soil, fertility, and slope differences throughout the whole plot of land. More information on the use of blocking and other forms of restricted randomization can be found in Box, Hunter, and Hunter (1978) and Steel and Torrie (1980).

Two critical issues concerning the conduct of the experiment are replication and randomization.

To summarize the features of a good experimental design, Table 3 (Snee, Hare, and Trout 1985) lists seven main characteristics. Some of these may seem difficult to achieve, but all are worthy goals. Without them an experiment may waste effort and yield erroneous conclusions. Most of these characteristics may seem to be little more than common sense; but, through its statistical and mathematical foundations, the statistical approach ensures that good experimental designs have them.

5. EXAMPLES OF DESIGNED EXPERIMENTATION

The environment in which the experiment is to be conducted dictates the design. Considerations include the objective, the types of factors to be studied and responses to be measured, the available resources (money, people, equipment), and deadlines for completing the results. Several examples in this section illustrate types of environments and the associated designs.

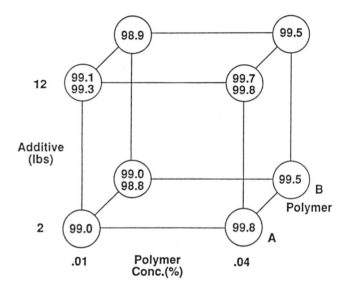

Figure 4: Outcomes of a 2^3 experiment, product purity in percent.

5.1 Studying Factors at Two Levels

Two-level factorial designs, in which each factor is studied at two levels, are the most common class of designs because they are easy to use and provide maximum information per experimental run. These designs yield very efficient measures of the linear effects of factors and their interactions. The Easter lily experiment discussed earlier used a 2×2 factorial design (Table 1).

The results of a $2 \times 2 \times 2$, or 2^3 (read "two to the three"), factorial design are shown in Figure 4. This experiment studied the effects of one qualitative and two quantitative factors, each at two levels, on the purity of a product. Two types of polymers (A and B) were in-

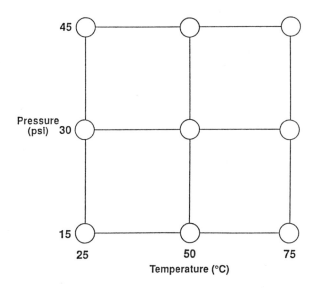

Figure 5: Plastic part process improvement, 3 × 3 factorial design.

volved in making the product. Each was studied at 2 concentrations (0.01, 0.04%). An additive was known to increase purity when used at a 2-pound rate. The question was whether higher purity percentages would result from using higher amounts of the additive, specifically 12 pounds.

The results of the 8 combinations in this 2^3 factorial design are shown in Figure 4. The runs were made in a random order. Three of the 8 combinations were replicated. As the experiment progressed, it was decided that further replication was not needed because the differences among the replicates were very small.

One beauty of the 2^3 design is that the results are easy to interpret. The linear effects of the factors can be studied by computing the changes in the response along the different edges of the cube (Figure 4). The linear effect of each factor (rounded to one decimal) is estimated four times as follows:

	Polymer Concentration	Polymer Type	Additives
	+0.8	−0.1	+0.2
	+0.6	−0.3	+0.0
	+0.6	−0.3	+0.0
	+0.6	−0.3	+0.0
Average	+0.65	−0.25	+ 0.05

We conclude that polymer concentration has a strong positive effect on product purity, a small but consistent difference separates the product purity levels of polymers A and B, and larger amounts of the additive did not further improve product purity. The fact that these factor effects were consistent on each of the four edges of the cube for each factor indicated that there were no interactions among the factors.

All these results were verified by a more formal statistical analysis appropriate to the 2^3 design. (We do not discuss that analysis here.) It was found that polymer A should be used along with 2 pounds of the additive. The polymer concentration would be selected based on the application and cost considerations, with 0.04% providing the best results. Further, it was learned that increasing additives should be avoided because, beyond the standard 2 pounds, they contribute nothing to purity except the cost of attaining it. Of course, it is possible that some additive level between 2 and 12 pounds results in even higher percentages of purity than the levels studied. The relationship between purity and additive level may not be linear. It is always wise to carry out additional experiments to confirm assumptions and results.

Many other issues surround the use of two-level factorial designs. When there are many factors, a class of factorial designs called fractional factorial designs can be used to help determine which factors or combinations of factors are most influential. Fractional factorial designs, used as screening designs, help direct attention to areas with the greatest potential gain. More information on this subject can be found in Box, Hunter, and Hunter (1978). Section 5.4 discusses a two-level screening experiment.

5.2 Studying Factors at Three Levels

Three-level factorial designs can also produce useful results efficiently, especially when the number of factors is small—say two or three. The following example, aimed at improving plastic parts, used a 3 × 3 factorial design in combination with a two-level design similar to that described in the previous section.

The defect rate of molded plastic parts was running around 1.7% and needed to be reduced. The parts were being produced using two different processes (old, new) and two different sources of raw material (A, B). Both processes could be operated over the same ranges of temperature (25–75°C) and pressure (15–45 psi). The operating conditions in use at the time were 50°C and 30 psi.

It was decided to study the effects of temperature and pressure separately for each process and type of raw material, using three levels of temperature (25, 50, 75°C) and three levels of pressure (15, 30, 45 psi). The result-

| | Raw Material A | | | Raw Material B | | |
| | Temperature (°C) | | | Temperature (°C) | | |
Pressure PSI	25	50	75	25	50	75
Old Process						
15	0	0	0.7	0.1	0	0
30	0.1	0	4.0	0.6	0	12.7
45	0	2.0	6.1	0.3	0	23.7
New Process						
15	0	0	0.4	0.1	0	0
30	0	0	0.6	0	0	2.4
45	0	13.9	2.4	0.4	0	19.7

Table 4: Reponses (percent defective parts) in the plastic part improvement experiment.

ing 9 combinations of temperature and pressure form a 3×3 factorial design (Figure 5). Four such experiments were run—one for each of the 4 combinations of process type (old, new) and type of raw material (A, B). The response was percent defective parts in a sample of 700 parts.

The data from this experiment are summarized in Table 4. We see immediately, without any statistical analysis, that no defective parts were observed at 50°C when the pressure was at 15 or 30 psi. These findings applied for both processes and both types of raw material.

The next step was to verify these results by running a second set of 3×3 factorial designs in the region of a temperature-pressure combination that produced no defective parts. In the verification test, temperature was studied at 30, 45, and 60°C with pressure at 10, 15, and 20 psi (Figure 6). No defective parts were observed at any of the 9 combinations of temperature and pressure for either process and both types of raw material lots. These were exciting results: a defect-free process had not been seen before.

The initial and verification tests had been run in the laboratory. It was now time to run a *plant test* to see whether the manufacturing process would behave similarly. The plant test was run at the center point of the 3×3 verification test (45°C, 15 psi). *No defects were observed in the plant test!!* A change in the standard operating procedures for the process was issued.

In monitoring over subsequent months, the process de-

fect rate dropped from its original 1.7% to 0.1%. This resulted in an annual savings of $500,000, to say nothing of the benefits, financial and otherwise, of increased customer satisfaction.

This example illustrates three key points. First the iterative nature of experimentation (Figure 3) is evident. We progressed from the initial test, to the verification test, and finally to the plant test. Second, it illustrates the power of the 3×3 factorial design and a key scientific principle: the need to establish the range of validity of the results through studying different processes and lots of raw material. Finally, it demonstrates that the factorial design does an effective job of sampling the experimental region. Hence, inspection of the data makes clear where the process should be operated. This is particularly true when only one response variable is of interest, as in this example (percent defective parts).

Exploring Response Surfaces

Response-surface methods are powerful tools for exploring experimental regions in which the factors are quantitative. Their graphical nature offers great advantages for easy interpretation and communication of results, as the following example illustrates.

A research effort to develop an assay method to make plasma ammonia measurements using an automated instrument (Humphries et al. 1979) involved three key variables known to affect the sensitivity of the instrument. These are buffer pH, enzyme concentration, and buffer molarity. It was decided to study the effects of these

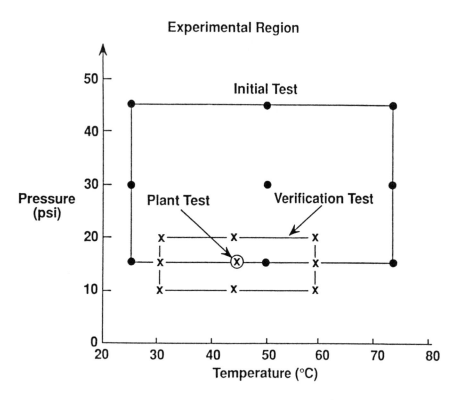

Figure 6: Plastic part process improvement verification run.

three variables over the following ranges:

Variable	Low	Middle	High
Buffer pH	7.25	7.45	7.65
Enzyme Concentration	90	125	160
Buffer Molarity	.04	.05	.06

One candidate design for this experiment, a $3 \times 3 \times 3$ factorial, would involve all possible combinations of the 3 factors, and thus would require 27 runs. As we will see later, using the concepts of response-surface methodology, we can satisfy the objective of the experiment using the 15 runs of the face-centered-cube design shown in Figure 7.

The experiment sought to determine the combination of these three variables that would maximize the sensitivity of the assay method. Assay sensitivity is measured by unit absorbance change per unit concentration of sample NH_4^+. Details of the experimental procedures and response measurement are discussed by Humphries et al. (1979). The strategy used was to collect data that would enable estimation of the coefficients (b's) in the

following second-order response-surface equation:

$$
\begin{aligned}
y = {} & b_0 + b_1 X_1 + b_2 X_2 + b_3 X_3 + b_{12} X_1 X_2 + b_{13} X_1 X_3 \\
& + b_{23} X_2 X_3 + b_{11} X_1^2 + b_{22} X_2^2 + b_{33} X_3^2
\end{aligned}
$$

where

$$
\begin{aligned}
y &= \text{sensitivity,} \\
X_1 &= (\text{pH} - 7.45)/0.2 \\
X_2 &= (\text{Enzyme} - 125)/35 \\
X_3 &= (\text{Molarity} - 0.05)/0.01.
\end{aligned}
$$

The face-centered-cube design in Figure 7 provides the data for efficient estimation of the coefficients in this equation. The 15 points in the face-centered-cube design consist of the 8 corners of the cube, the centers of the 6 faces of the cube, and the overall center of the cube.

Converting pH, enzyme concentration, and molarity to X_1, X_2, and X_3 centers and scales each of these factors. Each of the X's ranges from -1 to $+1$. Coefficients in the model with centered and scaled factors are directly comparable. For example, the coefficients of the linear terms (b_1, b_2, and b_3) relate to the change in response over the full range of the predictor variables (X_1, X_2,

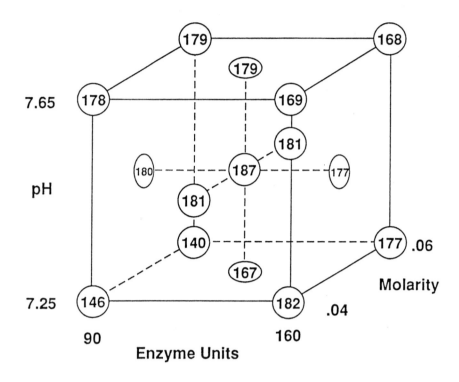

Figure 7: Three-factor face-centered-cube design

and X_3), and if, for example, b_1 were twice b_2, we could say that X_1 is twice as influential over its range as X_2 is over its range.

Fitting the second-order response-surface equation to the plasma ammonia results (Figure 7) by least-squares regression techniques (Draper and Smith 1981) produced the following regression coefficients:

$$b_0 = 184.4$$
$$b_1 = 3.7^{**} \quad b_{12} = -8.7^{**} \quad b_{11} = -7.7^{**}$$
$$b_2 = 4.0^{**} \quad b_{13} = 1.4 \quad b_{22} = -4.4$$
$$b_3 = -0.5 \quad b_{23} = 0.0 \quad b_{33} = -2.0$$

The asterisks (**) indicate that a regression coefficient is statistically significant at the .01 probability level, meaning that it is unlikely to have occurred by chance alone. We conclude that all variables except X_3 (buffer molarity) have a significant effect on sensitivity; hence, any buffer molarity in the range of 0.04 to 0.06 is acceptable. Similar conclusions are reached by looking at the trends in Figure 7. We can also conclude that the linear effect of pH over its range is nearly the same as the linear effect of "enzyme concentration" over its range. Although this conclusion is true, it is an over-simplification be-

cause the model also indicates that pH and enzymes act jointly (interact) to influence plasma ammonia, and pH has a curved effect.

Interpretation of the coefficients of such models can be confusing. Response-surface contour plots help provide clarification. In this example, we wanted to find maximum instrument sensitivity. Therefore, the contour plot shown in Figure 8 was constructed. It shows contours of predicted sensitivity as a function of buffer pH and enzyme concentration. Buffer molarity is held constant at 0.05.

The contour plot shows some important information. First, maximum sensitivity is reached in the region of pH = 7.45, enzyme concentration = 140. Also, the slanted and elongated contours indicate an interaction between buffer pH and enzyme concentration. This was also indicated by the statistically significant b_{12} regression coefficient ($b_{12} = -8.7$).

We notice that the response surface is relatively flat in the region of maximum sensitivity. Approximately one-third of the experimental region is predicted to give assay sensitivities of 181 or higher, which is within 98% of the maximum predicted sensitivity of 185. We con-

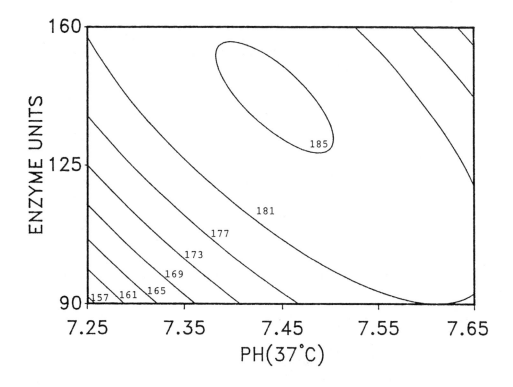

Figure 8: Ammonia assay study, sensitivity contour plot.

clude that this system is robust or rugged with respect to minor variation in buffer pH and enzyme concentration. This is a desirable characteristic for the system to have; otherwise, tight controls, and added expense, would be needed to maintain the sensitivity of the method. Other examples of the use of response-surface methodology in clinical chemistry and in developing trade-offs among multiple responses can be found in Rautela, Snee, and Miller (1979) and Myers (1985).

This example illustrates several key points. Using the concepts of response surface methodology, we were able to identify the region of maximum sensitivity and the behavior of method sensitivity in that region. This allowed us to conclude that a system centered in the region of maximum sensitivity would be rugged with respect to minor changes in all three variables studied.

The model that allowed us to draw these conclusions was constructed from only a small number of data points ($n = 15$). One-factor-at-a-time, or any other experimental strategy, not only would have failed to identify the region of maximum sensitivity (because OFAAT does not work well in the presence of interaction), but it also would have required considerably more experiments.

5.3 Blending and Product-Formulation Experiments

Mixture experiments differ from factorial experiments in that the factors must always sum to 1.0 or 100%. The proportion of any component is determined by the sum of the proportions of the remaining components. Thus the experimental region is a simplex. Mixture experiments are designed, analyzed, and interpreted using the techniques of response-surface methodology (Snee 1979). Designs are constructed using the entire simplex or a portion of it, depending on the existence of constraints on the compositions. The regression models incorporate the constraint on the sum of the components, and response-surface contours reveal the region of most favorable response.

A three-component example is given in Table 5 and Figure 9, which summarize the results of a mixture experiment designed to probe the effect of blending three vegetable oil components on the solid fat index taken at 50°F. The three components are stearine (which is vegetable oil solids of one kind of oil), vegetable oil (of a different oil type), and vegetable oil solids of yet a third type of oil (Hare 1974).

Run Number	Proportion of stearine	Proportion of vegetable oil	Proportion of vegetable oil solids	Response SF1–50°F
1	1	0	0	4.6
2	0	1	0	35.5
3	0	0	1	55.5
4	1/2	1/2	0	14.5
5	1/2	0	1/2	25.7
6	0	1/2	1/2	46.1
7	1/3	1/3	1/3	27.4
8	2/3	1/6	1/6	14.5
9	1/6	2/3	1/6	32.0
10	1/6	1/6	2/3	42.5

Table 5: Runs and responses in a three-factor mixture experiment.

Table 5 lists the solid fat index for each of ten experimental formulations spanning the full simplex. The model shown in Figure 9 was fitted using least-squares regression. Notice that it is similar in form to the model in the previous example, except that the constant term (b_0) is missing. This is characteristic of a certain class of mixture models. A term for $X_2 X_3$ is also missing because it offered no contribution to the ability of the model to fit the data.

Figure 9 illustrates the blending behavior of these three components. Its use is powerful because it predicts the solid fat content for all possible blends, and it points to trade-offs among blends. For example, many blends will yield an index of 30. Wise scientists will recommend the most cost-effective blend to achieve the desired result; and if they are really wise, they will take credit for the discovery!

Mixture experiments arise in the development of many different types of products, including gasoline, paint, cake mixes, animal feed, textile fiber blends, and aerosol propellants. Cornell (1990) gives detailed discussion of statistical approaches to the design of mixture experiments.

5.4 Experimenting with a Large Number of Factors

To scientists and engineers learning about experimental design techniques for the first time, the amount of work required to carry out a designed experiment may seem formidable. The examples discussed so far have all been relatively small, but what happens when the number of factors is large? For example, 10 factors each at only two levels, would yield 1024 treatment combinations. Clearly, this number of experimental runs is too large to be practical in most situations.

The following example illustrates the application of fractional-factorial designs, which allow us to experiment with many factors. The advantage of fractional-factorial designs is that they put experimenters in touch with the factors and interactions that are most likely to be important. Fractional-factorial designs don't waste precious experimental effort attempting to estimate interactions that are not likely to be important.

A study was initiated because of the perceived large variation in viscosity measurements produced by an analytical laboratory (Snee 1985b). This was a concern because viscosity was a key quality characteristic of a high-volume product. It was decided to conduct a ruggedness test (Youden and Steiner 1975, Wernimont 1977) of the measurement process to determine which variables, if any, were influencing the viscosity measurement.

Discussion of how the measurement process worked identified seven variables that might be important:

	Level	
Variable	Low (−)	High (+)
X_1 = Sample Preparation	M_1	M_2
X_2 = Moisture Measurement	Volume	Weight
X_3 = Mixing Speed (rpm)	800	1600
X_4 = Mixing Time (hrs)	0.5	3
X_5 = Healing Time (hrs)	1	2
X_6 = Spindle	S_1	S_2
X_7 = Protective Lid	Absent	Present

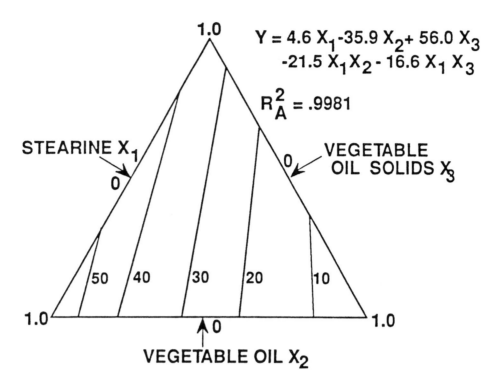

Figure 9: Contour plot for the outcome of a blending experiment. The axes go from the center of a side to the opposite vertex, as indicated by the arrows.

The sample was prepared by one of two methods (M_1, M_2) using moisture measurements made on either a volume or a weight basis. The sample was then put into a machine and mixed at a given speed (800 or 1600 rpm) for a specified time period (0.5 or 3 hrs) and allowed to "heal" for one or two hours. The levels of variables X_3, X_4, and X_5 reflected those used routinely by laboratory personnel making this test.

The mixer had two spindles (S_1, S_2) that were thought to be identical. The apparatus had a protective lid for safety purposes, and it was decided to run tests with and without the lid to see whether it had any effect.

The chemists in charge of the measurement process felt it was important to determine the effects of the first five variables. They were less enthusiastic about studying the effects of spindle (X_6) and protective lid (X_7), however, because they felt that these variables would not be important. They agreed to the seven-variable test when it was pointed out that including these two variables would not involve any additional runs beyond the sixteen required to test the first five variables.

The design used was a 16–run, 2^{7-3} fractional-factorial (Table 6). It permits estimation of all the main effects,

free of confusion with two-factor interactions. Many of the two-factor interactions are entangled with each other, so that, if such an interaction exists, we will be aware of it, but we will not know which one it is.

A few words about notation: 2^7 means that the design has 7 factors, each at 2 levels, for a total of 128 experimental treatment combinations. A 2^{7-1} design uses only half, or 64, of these experimental treatment combinations. A 2^{7-3} design uses one eighth of the treatment combinations. When we use a fraction of the possible set of experimental combinations, we sacrifice some information. This causes the "entanglement" of interactions. The trick in developing and applying fractional factorial designs is to minimize the entanglement of information thought likely to be important.

Results of runs 1–16, all made on samples from a common product source using a random test sequence, are shown in Table 6. Runs 4, 8, 12, and 16 were each done twice to measure the variation between repeated observations under the same operating conditions. Runs 1–16, including the replicates at runs 4, 8, 12, and 16, resulted in 20 tests (Table 6). If the measurement process were rugged (that is, if it were not influenced by various changes in operating conditions), then these results

Run	Test Sequence	X_1 Sample Prep	X_2 H$_2$O Meas	X_3 Mix Speed	X_4 Mix Time	X_5 Heal Time	X_6 Spindle	X_7 Lid	Y Viscosity (centipoise)
1	5	−	−	−	+	+	+	−	2220
2	4	+	−	−	−	−	+	+	2460
3	18	−	+	−	−	+	−	+	2904
4	19, 20	+	+	−	+	−	−	−	2464, 2348
5	7	−	−	+	+	−	−	+	3216
6	11	+	−	+	−	+	−	−	3772
7	12	−	+	+	−	−	+	−	2420
8	6, 13	+	+	+	+	+	+	+	2340, 2380
9	9	+	+	+	−	−	−	+	3376
10	3	−	+	+	+	+	−	−	3196
11	2	+	−	+	+	−	+	−	2380
12	1, 16	−	−	+	−	+	+	+	2800, 2700
13	14	+	+	−	−	+	+	−	2320
14	10	−	+	−	+	−	+	+	2080
15	15	+	−	−	+	+	−	+	2548
16	8, 17	−	−	−	−	−	−	−	2796, 2788
17		−	+	−	+	−	−	−	2384
18		−	+	+	+	−	−	−	2976
19		−	+	−	+	−	+	−	2180
20		−	+	+	+	−	+	−	2300

Table 6: A 2^{7-3} factorial design to study variation in viscosity measurements. (Runs 17–20 were made at least one week after runs 1–16.)

would differ only by the effects of random measurement error. Table 6 shows a nearly two-fold variation in the results, more than would be predicted by random variation.

The effects of the variables are summarized in Table 7 and shown graphically in a normal probability plot (Figure 10). For each effect (ordered from smallest to largest) this graphical display shows a "normal score," the average value of the corresponding ordered observation in samples of the same size from the standard normal distribution. When there are m effects, the expected value of the ith ordered observation is approximately $\Phi^{-1}(p_i)$, where Φ^{-1} is the inverse of the standard normal cumulative distribution function and

$$p_i = \frac{i - \frac{1}{2}}{m}, \qquad i = 1, 2, \ldots, m.$$

"Normal probability paper" (a special graph paper) has a uniform scale for the data (here the effects) and a probability scale for the values of p_i (usually given in percent). The transformation by Φ^{-1} is built into the grid.

In the present example the values of p_i for the 15 effects are 3.3%, 10%, 16.7%, ... , 96.7%. If experimental results simply exhibit random variation and are normally distributed, they will fall, within chance variation, along a straight line on normal probability paper. Conversely, if the effects do not resemble a straight line on normal probability paper, there is reason to suspect that effects far from the line are due to a cause other than chance.

Table 7 and Figure 10 indicate that the method of measuring moisture (X_2), mixing speed (X_3), mixing time (X_4), healing time (X_5), and spindle (X_6) are important variables. The effect associated with the entangled two-factor interactions, $X_1X_4 + X_3X_6 + X_5X_7$, is also large. An examination of the linear effects can help provide clues as to which variables are producing this effect. Among the individual variables, X_3 and X_6 have the largest effects. Often, variables with large linear effects also have significant interactions. Also, variables X_2 and X_7 have relatively small effects, suggesting that the X_1X_4 and X_5X_7 interactions may not be important, even though X_4 and X_5 have moderate effects.

Variable	Estimated Effect (High − Low)		
	Runs 1–16	Runs 1–20	Runs 1–20[1]
X_1 = Sample Prep Method	0	12	
X_2 = H$_2$0 Measurement	−140*	−145**	−147**
X_3 = Mixing Speed	343*	352**	345**
X_4 = Mixing Time	−343*	−312**	−307**
X_5 = Healing Time	124*	124**	123**
X_6 = Spindle	−684*	−624**	−608**
X_7 = Protective Lid	30	30	
X_8 = Experimental Error	33		
Interactions			
12 + 37 + 56	−46	−36	
13 + 27 + 46	−36	60	
14 + 36 + 57	−300		
14 + 57		−11	
36		−242**	−259**
15 + 26 + 47	−16	−18	
16 + 34 + 25	−30	−54	
17 + 23 + 45	−50	−52	
24 + 35 + 67	48	40	
Residual Std. Dev.	39[2]	73	81

*Effect is "large" on the normal probability plot
**Significant at the .01 probability level
[1]Nonsignificant terms deleted from the model
[2]Estimated from the 4 pairs of duplicate runs

Table 7: Estimated effects for the test variables and interactions in the viscosity measurement study.

It was decided to determine the source of the interaction by running a second experiment in which the levels of X_3 and X_6 were varied using a 2^2 design and the other 5 variables were held fixed as indicated in Table 6. The results of runs 17–20 (Table 6) were:

Spindle	Speed	
	800	1600
S_1	2384	2976
S_2	2180	2300

Figure 11 shows that these results are in good agreement with those predicted from runs 1–16. That is, both mixing speed and choice of spindle, and their interaction, influence the viscosity measurement. Spindle 1 gives higher results than Spindle 2, and the effect of mixing speed is greater for Spindle 1 than for Spindle 2. Analysis of the results of all 20 runs confirmed the importance of the interaction between spindle and speed.

It is clear from the results of this test that the viscosity measurement process is not rugged and that the levels of five of the seven variables have to be tightly controlled to maintain good measurement precision. All of this was learned at the expense of only 24 tests (Table 6).

Also, the two spindles, which were thought to be "identical," produced the largest difference in the study. In general, experimenters should not rule out the importance of any variable unless they have data to support their position. "In God we trust; others must have data."

6. THE IMPORTANCE OF GRAPHICS

We need all the help we can get when grappling with the problems of science and technology. Statistics and mathematics appeal naturally to the analytically-oriented left side of the brain. Through the use of graphical techniques we can engage the right side of the brain to help

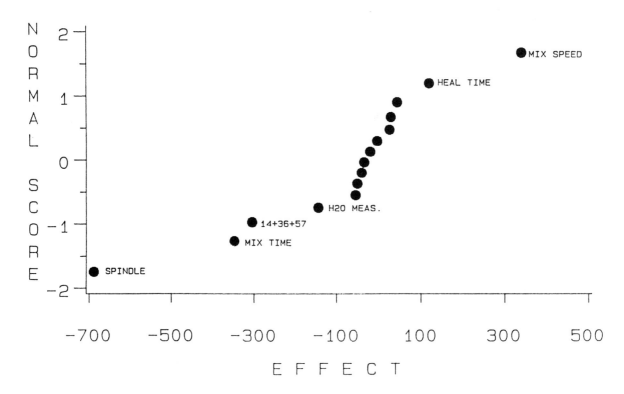

Figure 10: Normal probability plot of estimated effects for the viscosity measurement study.

us understand issues and results and communicate these to others. Graphical techniques enable us to

- better understand the region of experimentation;

- construct designs that adequately sample the region;

- analyze the data and detect abnormal results that may be distorting the statistical analysis and resulting interpretation;

- interpret the results of our statistical models through the use of plots of raw data and averages, interaction plots, normal probability plots, and response-surface contour plots; and

- communicate our findings to other scientists, engineers, and decision makers who will take actions based on them.

The old saying that "a picture is worth a thousand words" is equally true in the statistical approach to design of experiments.

In reflecting on the use of graphics in experimentation it is helpful to review the examples presented and consider how the graphics help us understand the example.

It is our firm conviction that statistical design of experiments, or any other aspect of statistics, cannot effectively solve problems and generate new knowledge without the integral, ubiquitous use of graphical displays (Snee and Pfeifer 1983, Tufte 1982, Cleveland 1985).

7. FOUNDATIONS OF STATISTICAL DESIGN OF EXPERIMENTS

Successful design of experiments requires knowledge of

- the scientific method,

- the field of science or technology in which the experimentation is being done,

- statistics,

- mathematics,

- computer science, and

- behavioral science.

In-depth knowledge of one, or even a few, of these fields is not sufficient to use statistical design of experiments successfully. To realize the full power of the methodology, one must have a working knowledge, not necessarily

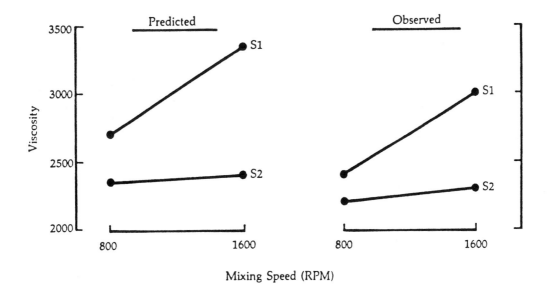

Figure 11: Plot of observed and predicted interaction between mixing speed and spindle in the viscosity measurement study.

obtained by formal education, in all of the fields (Snee 1983).

The most useful fields of mathematics are solid and analytic geometry, calculus, matrix theory, and linear and quadratic programming. It should be recognized that many of the methods of statistical design of experiments can be and are successfully used with only the knowledge of simple arithmetic operations. However, the more knowledge one has of these fields of mathematics, the more effective one will be in using the approach. The theory of Galois fields, for example, has been used for constructing designs (Finney 1960). Understanding optimal design theory requires a detailed knowledge of matrix theory (Federov 1972). It is interesting to note that both authors of this chapter have undergraduate degrees in mathematics.

Knowledge of science, technology, and the scientific method are needed so that one can communicate effectively with the team doing the experimentation and understand how the statistical approach fits into the process of experimentation.

Knowledge of computer science and how to use computers to solve problems is needed in design (Snee 1985a), analysis, and interpretation. We know of experiments in which computers constructed the design, directed the experimental apparatus, collected the data, and did the analysis. All of the computer software for this was, of course, constructed by humans, but one got the feeling that the data were "untouched by human hands."

Knowledge of human behavior is critical to seeing that experiments are designed cooperatively and correctly and that the changes suggested by the results are, in fact, implemented by the organization. Statisticians can benefit by obtaining feedback on their consulting ability and interpersonal skills (McCulloch et al. 1985). We all have a natural resistance to change. It can be overcome if we understand the emotional needs of the people we work with and how they are likely to interact with one another.

It is also critical to understand how the proposed changes will affect the different groups and how to get the support of all affected groups, particularly the leadership of the organization. Those who practice statistical consulting need to know more about principles of communication and teaching. A working knowledge of the principles of behavioral science is extremely valuable to mathematicians and statisticians and to other scientists and engineers.

8. IMPROVED QUALITY OF EXPERIMENTATION IS OUR GOAL

The statistical approach to design of experiments applies to all types of experimental studies and has been found to speed the progress of experimental programs. The statistical approach integrates well with the scientific method and iterative nature of experimentation and, as a result, works at the interface between statistics and science.

Students of all ages (from high school through on-the-job training) and backgrounds enjoy its study, particularly when they have the opportunity to perform their own experiments (Hunter 1977). It's also an interesting and enjoyable subject to teach.

Mathematics is one of the key (but not the only) foundations of statistical design of experiments (Snee 1988). Mathematics departments that offer courses and programs in statistics are obligated to expose their students, and students from other disciplines, to this important field of study.

The result of this expanded teaching of the statistical approach to the design of experiments will be higher-quality experimentation, more effective and efficient use of personnel and equipment, and broader use of mathematics. All parts of society will benefit in the process.

Ronald D. Snee is Leader of Continuous Improvement Development for Du Pont's Corporate Continuous Improvement Process. He joined Du Pont in 1968 after receiving a B.A. degree in mathematics from Washington and Jefferson College and M.S. and Ph.D. degrees in statistics from Rutgers University. His previous assignments at Du Pont included management of the development and implementation of continuous process improvement systems for project engineering, direction of Du Pont internal consultants, and service as Senior Consultant, Corporate R&D Planning. He also serves as an adjunct professor at the University of Delaware.

Dr. Snee has published extensively on statistics, quality, and applications. He has received the American Society for Quality Control's Brumbaugh, Frank Wilcoxon, Jack Youden, Ellis R. Ott, Shewell, and William G. Hunter awards. In 1986 he received the Shewhart Medal, ASQC's highest award. He is a fellow of the American Statistical Association, the American Society for Quality Control, and the American Association for the Advancement of Science.

Lynne B. Hare directs the Technical Services Department at Thomas J. Lipton, where he has been instrumental in introducing the principles of statistical thinking, data-driven decisions, and total quality management. A career statistician, he holds an A.B. degree from the Colorado College and M.S. and Ph.D. degrees from Rutgers University. His research interests are quality management, statistical applications in quality and productivity, and mixture designs, especially as they apply to food formulation and processing.

He is a fellow of the American Society for Quality Control, past-chair of its Statistics Division, and a member of its Standing Review Board. He also serves on the *Technometrics* Management Committee and is active in the American Statistical Association.

REFERENCES

Box, G.E.P., Hunter, W.G. and Hunter, J.S. (1978), *Statistics for Experimenters*, New York: John Wiley.

Business Week (1987), "The Push For Quality," *Business Week*, June 8, 1987, 130–143.

Cleveland, W.S. (1985), *The Elements of Graphing Data*, Monterey, CA: Wadsworth.

Cornell, J.A. (1990), *Experiments With Mixtures*, second edition, New York: John Wiley.

Deming, W.E. (1986), *Out of the Crisis*, Cambridge, MA: MIT Center for Advanced Engineering Study.

Draper, N.R. and Smith, H. (1981), *Applied Regression Analysis*, second edition, New York: John Wiley.

Federov, V.V. (1972), *Theory of Optimal Experiments*, New York: Academic Press.

Finney, D.J. (1960), *The Theory of Experimental Design*, Chicago: The University of Chicago Press.

Fisher, R.A. (1935), *The Design of Experiments*, Edinburgh: Oliver and Boyd.

Hare, L.B. (1974), "Mixture Designs Applied to Food Formulation," *Food Technology*, **28**, 50–56, 62.

Humphries, B.A., Melnychuk, M., Snee, R.D. and Donegun, E.J. (1979), "Automated Enzymatic Assay for Plasma Ammonia," *Clinical Chemistry*, **25**, 26–30.

Hunter, W.G. (1977), "Some Ideas About Teaching Design of Experiments with 25 Examples of Experiments Conducted by Students," *The American Statistician*, **31**, 12–17.

McCulloch, C.E., Boroto, D.R., Meeter, D., Pollard, R. and Zahn, D.A. (1985), "An Expanded Approach to Educating Statistical Consultants," *The American Statistician*, **39**, 159–167.

Myers, G.C., Jr. (1985), "Use of Response Surface Methodology in Clinical Chemistry," in *Experiments in Industry*, eds. R.D. Snee, L.B. Hare and J.R. Trout, 59–74, Milwaukee, WI: American Society for Quality Control.

Ott, E.R. (1975), *Process Quality Control*, New York: McGraw-Hill.

Penzias, A. (1989), "Editorial—Teaching Statistics to Engineers," *Science*, **244**, 1025.

Rautela, G.S., Snee, R.D. and Miller, W.K. (1979), "Response Surface Co-optimization of Reaction Conditions in Clinical Chemical Methods," *Clinical Chemistry*, **25**, 1954–1964.

Snee, R.D. (1979), "Experimenting with Mixtures," *CHEMTECH*, **9**, 702–710.

—— (1983), "Statistics in Industry," in *Encyclopedia of Statistical Sciences*, eds. S. Kotz, N.L. Johnson, and C.B. Read, 69–73, New York: John Wiley.

—— (1985a), "Computer Aided Design of Experiments—Some Practical Experiences," *Journal of Quality Technology*, **17**, 222–236.

—— (1985b), "Experimenting With a Large Number of Variables," in *Experiments in Industry—Design Analysis and Interpretation of Results*, eds. R.D. Snee, L.B. Hare and J.R. Trout, 25–35, Milwaukee, WI: American Society for Quality Control.

—— (1988), "Mathematics Is Only One Tool That Statisticians Use," *College Mathematics Journal*, **19**, 30–32.

—— (1990), "Statistical Thinking and Its Contribution to Total Quality," *The American Statistician*, **44**, 116–121.

Snee, R.D., Hare, L.B. and Trout, J.R., (eds.) (1985), *Experiments in Industry—Design, Analysis and Interpretation of Results*, Milwaukee, WI: American Society for Quality Control.

Snee, R.D. and Pfeifer, C.G. (1983), "Graphical Representation of Data," in *Encyclopedia of Statistical Sciences*, Volume 3, eds. S. Kotz, N.L. Johnson and C.B. Read, 488–511, New York: John Wiley.

Steel, R.G.D. and Torrie, J.H. (1980), *Principles and Procedures of Statistics—A Biometrical Approach*, second edition, New York: McGraw-Hill.

Tufte, E.R. (1982), *The Visual Display of Quantitative Information*, Chesire, CT: Graphics Press.

Wernimont, G. (1977), "A Ruggedness Evaluation of Test Procedures," *ASTM Standardization News*, **5**, 61–64.

Youden, W.J. and Steiner, E.H. (1975), *Statistical Manual of the Association of Official Analytical Chemists*, Arlington, VA: Association of Official Analytical Chemists.

Chapter 6

What Is Probability?

Glenn Shafer
University of Kansas

1. INTRODUCTION

What is probability? What does it mean to say that the probability of an event is 75%? Is this the frequency with which the event happens? Is it the degree to which we should believe it will happen or has happened? Is it the degree to which some particular person believes it will happen? These questions have been debated for several hundred years. Anyone who teaches statistics should have some sense of how this debate has gone and some respect for the different viewpoints that have been expressed. Each seems to have its germ of truth.

This chapter introduces the debate to those who are not familiar with it. It also sketches a way of reconciling the different viewpoints and draws some lessons for the teacher of probability and statistics.

It is conventional to say that mathematical probability theory has a number of different interpretations. The same mathematical rules (Kolmogorov's axioms and definitions) are obeyed by degrees of belief, by frequencies, and by degrees of evidential support. We can study these rules for their own sake (this is pure mathematics), or we can adopt one of the interpretations and put the rules to use (this is statistics or applied probability). Section 2 explores this conventional formulation. It reviews Kolmogorov's axiomatization and the three standard interpretations of this axiomatization: the belief interpretation, the frequency interpretation, and the support interpretation. Each interpretation, as we shall see, has its appeal and its difficulties.

Section 3 very briefly reviews how the three standard interpretations handle statistical inference. This reveals that they are not as distinct as they first appear. The

belief and support interpretations both use Bayesian inference. Moreover, they use Bayesian inference to find beliefs (or degrees of support) about frequentist probabilities. Proponents of the frequency interpretation reject Bayesian inference in most cases, but they too end up interpreting certain probabilities as beliefs about frequentist probabilities.

It is conventional to say that mathematical probability theory has a number of different interpretations. The same mathematical rules (Kolmogorov's axioms and definitions) are obeyed by degrees of belief, by frequencies, and by degrees of evidential support. We can study these rules for their own sake (this is pure mathematics), or we can adopt one of the interpretations and put the rules to use (this is statistics or applied probability).

Perhaps frequency, degree of belief, and degree of support are *not* merely three distinct interpretations of the same set of axioms, unrelated except for the coincidence that they follow the same mathematical rules. Perhaps they are more entangled than this—so entangled that it is more accurate to say that they are aspects of a single complex idea. This is the thesis of Section 4, which argues that probability is a complex idea, one that draws together ideas about fair price, rational belief, and knowledge of the long run.

Section 5, in conclusion, draws some lessons for teaching.

The most important lesson is humility. Whenever we tell students, "This is what probability really means," we are wrong. Probability means many things.

2. THREE INTERPRETATIONS OF KOLMOGOROV'S AXIOMS

For the pure mathematician of probability, the axioms and definitions that A.N. Kolmogorov published in 1933 are inseparable from his demonstration that they could be used as a rigorous basis for the study of infinite sequences of random variables. Here, however, we are not interested in infinity. We are interested instead in the implications of Kolmogorov's axiomatization for the meaning of probability, and for this purpose we can work with a finite sample space.

Suppose Ω is a finite sample space. Call the subsets of Ω events. Suppose a probability $P(A)$ is assigned to each event A. Under these assumptions, Kolmogorov's axioms are equivalent to the following slightly long-winded list of axioms:

> Axiom 1. For each $A, 0 \leq P(A) \leq 1$.
>
> Axiom 2. If A is impossible, then $P(A) = 0$.
>
> Axiom 3. If A is certain, then $P(A) = 1$.
>
> Axiom 4. If A and B are incompatible, then $P(A \cup B) = P(A) + P(B)$.

Here "A is impossible" means that $A = \emptyset$, "A is certain" means that $A = \Omega$, and "A and B are incompatible" means that $A \cap B = \emptyset$.

We could make this list of axioms more concise. We could omit Axiom 3, for example, because it follows from Axioms 2 and 4. But we are interested here in the meaning and justification of the axioms and definitions, not in the most concise way of stating them.

Kolmogorov's axiomatization of probability consists of his axioms together with several definitions. If $P(A) > 0$, then we call

$$P(B|A) = \frac{P(A \cap B)}{P(A)} \qquad (1)$$

the *conditional probability* of B given A. We say that A and B are *independent* if $P(B|A) = P(B)$. We call a real-valued function X on Ω a *random variable*. We set

$$E(X) = \sum_{i \varepsilon \Omega} X(i) \cdot P(\{i\}), \qquad (2)$$

and we call $E(X)$ the expected value of X.

We will review in turn three standard interpretations of Kolmogorov's axiomatization. The first interpretation takes $P(A)$ to be a person's degree of belief that A will happen. The second takes $P(A)$ to be the frequency with which A happens. The third takes $P(A)$ to be the degree to which the evidence supports A's happening, or the degree to which it is rational to believe that A will happen. Savage (1972) called these the personalistic, objectivistic, and necessary interpretations. They have also been discussed by Nagel (1939), Kyburg and Smokler (1980), Barnett (1982), and many others.

2.1 Belief

The belief interpretation is really a betting interpretation. When a person says her probability for A is 75%, we assume that she will back this up by betting on A and giving 3 to 1 odds. We also assume that she is equally willing to take the other side of such a bet.

Let us review what giving 3 to 1 odds means. It means putting 75 cents on A if the other person puts 25 cents against A. You will lose the 75 cents to the other person if A does not happen, but you will win the other person's 25 cents if A does happen. In effect, you are paying 75 cents for a ticket that returns \$1 if A happens. Taking the other side of the bet means paying 25 cents for a ticket that returns \$1 if A does not happen.

In a nutshell, then, your probability for A is the price you will pay for a \$1 ticket on A. You will pay half as much for a 50-cent ticket on A, and twice as much for a \$2 ticket on A.

Why should such prices satisfy Kolmogorov's axioms? And what is the point, if this is what we mean by probability, of defining conditional probability, independence, and expected value in the way Kolmogorov does?

We can argue persuasively for the four axioms. Consider Axiom 4, for example. Suppose A and B are incompatible, your probability for A is 40%, and your probability for B is 20%. Then you would pay 40 cents for a ticket that pays \$1 if A happens, and you would pay 20 cents for a ticket that pays \$1 if B happens. If you buy both tickets, then in effect you are paying 60 cents for a ticket that pays \$1 if $A \cup B$ happens. So your probability for $A \cup B$ must be 60%. Thus $P(A \cup B) = P(A) + P(B)$.

This argument can be elaborated in various ways. One

way is to imagine that you post odds for every event and allow another person to choose what bets to make with you at those odds and which side of the bet to take in each case. In this case, you must satisfy Axioms 1 to 4 in order to keep the person from choosing bets in such a way that she will make money for certain, no matter how the events come out. This is sometimes called the Dutch-book argument.

Conditional probability, in the belief or betting interpretation, is the same as probability—it is probability under new circumstances. Suppose you know that A will happen or fail before B. Then the conditional probability $P(B|A)$ is the degree to which you will believe in B right after A happens—if it happens. In betting terms, it is the amount you will be willing to pay right after A happens for a $1 ticket on B.

This makes it easy to explain the definition of independence. Saying that A and B are independent means that the happening or failing of A will not change your probability for B.

Why should conditional probabilities obey formula (1)? Suppose your probability for A is 60%, and your conditional probability for B is 50%. Then you are willing to pay 30 cents for a 50-cent ticket on A, and if it does happen, then you are willing to pay the 50 cents you have just won for a $1 ticket on B. If you plan to spend the 50 cents in this way if you win it, then when you pay your 30 cents, you are in effect buying a ticket that pays $1 if A and B both happen. Thus your probability for $A \cap B$ is 30%. These probabilities thus obey $P(A \cap B) = P(A) \cdot P(B|A)$. This is called the rule of compound probability, and it is essentially equivalent to (1).

We are interpreting the probability $P(A)$ as the amount you are willing to pay for a $1 ticket on A. This ticket pays $1 if A happens and $0 if A does not happen. We can interpret $E(X)$, as given by (2), as the price you are willing to pay for a more complicated ticket X. This ticket pays $X(i)$, where i is the outcome. If $P(\{i\})$ is your probability for i, then you are willing to pay $[X(i) \cdot P(\{i\})]$ for a ticket that pays $X(i)$ if i happens and $0 otherwise. If you buy a ticket like this for each i, then in effect you have bought the ticket X. The amount you have spent is the sum on the right-hand side of (2).

There are some obvious objections to these arguments. First, we may be assuming too much when we assume that a person is willing to set odds on each event and bet

on either side of these odds. You can imagine a person being more cautious. She might require you to offer her more than even odds, for example, before she would bet either for or against a particular event. If we allow her to behave in this way, and we still think of the greatest odds she is willing to give on an event as measuring her degree of belief or probability for the event, then we get probabilities that may not add as required by Axiom 4. Such non-additive probabilities have been studied by many authors, including Walley (1991).

Are Kolmogorov's axioms supposed to tell us how a person's degrees of belief *should* fit together? Or are they supposed to describe how people actually behave when given opportunities to bet or when facing other decisions under uncertainty?

The argument for the rule of compound probability also involves some strong assumptions. We assume that the events A and B happen or fail in sequence and that we will know as soon as A happens or fails. We also assume that our probability for B right afterward is well-defined; it does not vary with other circumstances involved in A's happening or failing. Many authors, especially de Finetti (1974, 1975), have tried to relax these assumptions, but then the argument becomes less persuasive (Shafer 1985).

Another point of controversy is whether the belief interpretation is normative or descriptive. Are Kolmogorov's axioms supposed to tell us how a person's degrees of belief *should* fit together? Or are they supposed to describe how people actually behave when given opportunities to bet or when facing other decisions under uncertainty? Most statisticians who subscribe to the belief interpretation say that Kolmogorov's axioms are primarily normative. Whether people conform to these axioms in everyday life is not important to the work of a statistician. Outside statistics, however, the value of the belief interpretation as a descriptive theory is widely debated. Psychologists have given many examples of ways that people do not conform to the axioms in their judgments of probability and in their decisions (Tversky and Kahneman 1986), yet a good deal of modern economic theory assumes that the axioms have some descriptive (or at least predictive) validity (Diamond and Rothschild 1978).

2.2 Frequency

According to the frequency interpretation, the probability of an event is the long-run frequency with which the event occurs in a certain experimental setup or in a certain population. This frequency is a fact about the experimental setup or the population, a fact independent of any person's beliefs.

Suppose we perform a certain experiment n times, under identical conditions, and suppose a certain event A happens k times. Then the *relative frequency* of A is

$$\frac{k}{n}.$$

Perhaps there is a particular number p toward which this ratio always converges as n increases. If so, then p is the probability of A in the frequency interpretation.

Many people object to the acknowledged narrow scope of application of the frequency interpretation. Many events for which we would like to have probabilities clearly do not have probabilities in the frequency sense.

The frequency interpretation is less widely applicable than the belief interpretation. A person can have beliefs about any event, but the frequency interpretation applies only when a well-defined experiment can be repeated and the ratio (3) always converges to the same number.

The frequency interpretation makes Kolmogorov's axioms easy to justify. The axioms obviously hold for the relative frequencies given by (3). The relative frequency of A is always between zero and one. It is zero if A is impossible and never happens, and it is one if A is certain and always happens. If A and B are incompatible events, A happens k_A times and B happens k_B times, then $A \cup B$ happens $k_A + k_B$ times, and hence the relative frequencies add. The probabilities of the events are the limits of these relative frequencies as n, the number of trials, increases. Because the axioms hold for the relative frequencies, they hold for their limits as well.

The conditional probability of B given A, in the frequency interpretation, is the limit of the relative frequency of B in those trials in which A also happens.

Formula (1) follows directly from this definition. Independence means that B happens overall with the same relative frequency as it happens in the trials in which A happens. The expected value of a random variable X is simply the long-run average value of X in many trials.

Many people object to the acknowledged narrow scope of application of the frequency interpretation. Many events for which we would like to have probabilities clearly do not have probabilities in the frequency sense.

We can also question whether the frequency interpretation gives an adequate motivation for the definitions of conditional probability and independence. Why, for example, should we care about the relative frequency of B in the trials where A happens? If we find out that A has happened, then this relative frequency does seem more relevant than the overall relative frequency of B as a guide to whether B will happen. But saying this seems to take us out of the domain of objective frequencies into the domain of belief.

It is also odd, if we begin with frequency as the definition of probability, that we should then expend great effort to prove the law of large numbers—the theorem that the probability of an event will almost certainly be approximated by the event's relative frequency. This was seen as a real problem by the frequentists of the nineteenth century (Porter 1986). But most frequentists nowadays take a more relaxed attitude. Frequency is the definition of probability in practice, they say, but it is convenient in the purely mathematical theory to take probability as a primitive idea and to prove the law of large numbers as a theorem.

Here is a related puzzle. In order to prove the law of large numbers for an event A in our experiment, we must consider a compound experiment, consisting of n trials, say, of the original experiment. We assign probabilities to the possible outcomes of this sequence of experiments, using the probabilities for the original experiment and assuming independence of the trials. Then we choose some number ε, we consider the event B that the relative frequency of the event A in the n trials will be within ε of $P(A)$, and we prove that the probability of B is high. This gives us a frequency interpretation of $P(A)$. But what about the probability of B, and the probabilities of all the other events that can be defined in terms of the n trials? Do they have frequency interpretations? No problem, say many frequentists. We simply consider a yet larger experiment, involving sequences of sequences of trials (Cramér 1946).

The preceding objections have not troubled twentieth-century frequentists, but they have taken a more concrete problem very seriously. This is the problem that probability theory seems to require more than mere convergence of relative frequencies to limits. The convergence must take place at a certain tempo. Yet the frequency interpretation does not impose this. Thus mere frequency does not seem adequate, as a model in the formal sense, for probability theory.

Richard von Mises, in the 1920s and 1930s, proposed that we model probability theory not merely with frequencies but with whole sequences of outcomes. He coined the name "Kollektiv" for a sequence of outcomes whose relative frequencies converge in the manner expected of a random sequence in probability theory. Von Mises's ideas were developed in the 1930s by Jean Ville and Abraham Wald, who showed that it is possible to find sequences of outcomes that satisfy any countable number of the properties that we would expect from a random sequence (Martin-Löf 1969).

During the last three decades, von Mises's ideas have been developed further in terms of the complexity of a sequence, which can be defined as the length of a computer program needed to generate the sequence. A number of mathematicians, including Kolmogorov, have shown that sequences that come close to being maximally complex tend to have the properties we expect from a random sequence (Cover et al. 1989).

2.3 Support

According to the support interpretation, probability is rational degree of belief. The probability $P(A)$ of an event A is the degree to which we should believe A will happen—the degree to which our evidence supports A's happening.

What reason do we have for thinking that there is a precise numerical degree to which our evidence supports A's happening? Twentieth-century proponents of the support interpretation concede that it is difficult to measure degrees of support, but they are convinced that evidence does give support for beliefs. This support may be qualitative rather than quantitative, but it follows certain rules nevertheless, and we can make it quantitative by adopting certain conventions. Kolmogorov's axioms and definitions follow from these qualitative rules and conventions.

One of the most basic qualitative rules advanced by proponents of the support interpretation is the rule that if A and B are incompatible, then the degree of support for $A \cup B$ is completely determined by the degrees of support for A and B. To this we may add that it is an increasing function of these two degrees of support; the more support there is for A or for B, the more there is for $A \cup B$. Once we accept these qualitative rules, the numerical rule given by Axiom 4 appears to be a harmless convention (Jeffreys 1961). In fact, it can be derived using a few regularity conditions (Cox 1961, Schrödinger 1947).

Conditional probability and the rule of compound probability can be dealt with similarly. We define conditional probability by saying that $P(B|A)$ is the degree of support for B provided by our present evidence together with the further knowledge that A has happened. We formulate the qualitative rule that the degree of support for $A \cap B$ is completely determined by the degree of support for A based on the current evidence, together with the degree of support for B based on that evidence and knowledge of A. We add that it is an increasing function of both, and we then present the rule of compound probability as a convention or derive it using additional regularity conditions.

What reason do we have for thinking that there is a precise numerical degree to which our evidence supports A's happening? Twentieth-century proponents of the support interpretation concede that it is difficult to measure degrees of support, but they are convinced that evidence does give support for beliefs.

It is easy to raise objections to this approach. To begin with, we can question the qualitative rules. Why should the degree of support $P(A \cup B)$ depend only on the degrees of support $P(A)$ and $P(B)$, and not on other aspects of these two events or other aspects of the evidence? There does not seem to be any argument for this qualitative principle, aside from the fact that the familiar numerical rule satisfies it. In some alternative theories (e.g., Shafer 1976) the principle is not satisfied.

Even if we accept the existence of well-defined degrees of support based on our current evidence, we can question whether conditional degrees of support exist. Because we might learn that A is true in many different ways, it may not be appropriate to talk without qualification about the support for B based on our current evidence together with knowledge of A (Shafer 1985).

3. THE THREE INTERPRETATIONS IN PRACTICE

We have been discussing the three interpretations of probability as if they were completely unrelated—as if Kolmogorov's axioms and definitions were all they had in common. In fact, the three interpretations are thoroughly entangled. They are entangled historically, conceptually, and practically. In this section, we look at the entanglement in statistical practice.

In statistical practice, proponents of all three interpretations are interested in both frequencies and degrees of belief. All three groups make inferences about frequencies, and they all use probabilities, in one way or another, to express these inferences. They disagree about how to make inferences; proponents of the belief interpretation and the support interpretation use Bayesian methods, whereas proponents of the frequency interpretation use sampling-theory methods. But in both cases, the inferences are about frequentist probabilities. There is also disagreement about how to express the inferences; sampling-theory methods use probabilities with subtle frequency interpretations, while Bayesian methods use probabilities that are labeled outright as degrees of belief or degrees of support. But in practice even the probabilities produced by sampling-theory methods are interpreted as degrees of belief.

Bayesian and sampling-theory inference are discussed in more detail in Chapters 1 and 7 of this volume. Discussions that emphasize the differences between the two approaches include Efron (1978) and Barnett (1982).

3.1 Bayesian Inference

Suppose we flip a coin 10 times, and we get the sequence $HHTTHTHHHH$—7 heads and 3 tails altogether. What should we think about the true probability of heads?

If we write p for the true probability of heads, then the probability of the sequence $HHTTHTHHHH$ is

$$p^7(1-p)^3. \tag{3}$$

This is graphed as a function of p in Figure 1. It seems

reasonable to take this function as a measure of how much we should believe different values of p. Thomas Bayes and Pierre-Simon Laplace, two eighteenth-century students of probability, suggested that we use it to find probabilities for p. We multiply (4) by a constant so that it will integrate to one, and we use the result as a probability density:

$$f(p) = 1320p^7(1-p)^3. \tag{4}$$

Using this probability density, we can give probabilities for any interval of values for p. The probability of p being between 0.54 and 0.86, for example, is 77%.

The probability density given by (5) is called the *posterior density* for p. The function given by (4) is called the *likelihood function*. We have simply multiplied (4) by a constant to get (5), but we can incorporate into the process a probability density $g(p)$ based on other evidence. We call $g(p)$ the *prior density*, and we take the posterior density to be proportional to the product of $g(p)$ and the likelihood function. Thus formula (5) gives the posterior density for our problem only in the case where the prior $g(p)$ is uniform (i.e., where $g(p) = 1$ for all $p, 0 \leq p < 1$).

This approach to statistical inference was made popular by Laplace, and in the nineteenth century it was called the method of inverse probability. Today we call it Bayesian inference, and we base it on what we anachronistically call Bayes's theorem. Bayes's theorem says that if $A_1, A_2, \ldots, A_n$ are incompatible hypotheses, one of which must be true, then

$$P(A_i|B) = KP(A_i) \cdot P(B|A_i),$$

where the constant K does not depend on A. Here $P(A_i|B)$ is the posterior, $P(A_i)$ is the prior, and $P(B|A_i)$ is the likelihood. This theorem is easy to prove if we accept Kolmogorov's axioms as a starting point, but it is conceptually troublesome, because it involves conditional probabilities in two directions (the probability of A_i given B and the probability of B given A_i), whereas the justification of conditional probability that we reviewed in Section 2.1 relies on a single sequence of events, with the assumption that both the events themselves and our knowledge of them unfold together in that sequence.

Both the support and the belief interpretations use Bayesian inference, but they differ in their interpretation of the prior probabilities. Proponents of the belief interpretation regard prior probabilities as personal beliefs. Proponents of the support interpretation try to

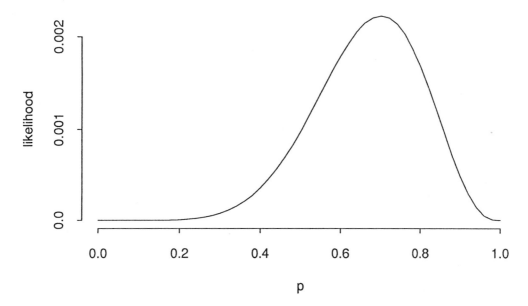

Figure 1: The likelihood function for p resulting from 7 heads and 3 tails

find objective grounds for choosing a prior distribution. In the case of the coin, for example, they regard the uniform prior density as an expression of ignorance. The support interpretation is rejected by most statisticians, because the case for objective priors is a confused one. But the belief interpretation has its own problem here. This lies in the apparently objective and frequentist nature of the true probability p. Are we giving probability a belief interpretation if we interpret the prior and posterior probabilities as beliefs but interpret the "true unknown probability p" itself as an objective property of the coin?

De Finetti (1974, 1975) has argued that the apparently frequentist p is merely a way of talking; behind it lie purely subjective ideas about the symmetry of our beliefs about a long sequence of successive flips of the coin. Nonetheless, the workaday world of Bayesian statistics seems to accept a dual interpretation of probability. In practice, Bayesians accept models that hypothesize frequentist probabilities. They differ from the frequentists only in that they use probabilities interpreted as beliefs in order to make inferences about the probabilities interpreted as frequencies.

3.2 Sampling-Theory Inference

Consider again the problem of making judgments about the true probability p after observing 10 tosses. The frequentist approach considers the frequency properties of different ways of making such judgments.

Suppose we write X for the number of heads in 10 tosses, and we say that we are going to estimate p by $X/10$. The expected value of this estimator is p, and its standard deviation is $\sqrt{p(1-p)/10}$, which is equal to at most 0.16. By the central limit theorem, we expect that the estimator will be within one standard deviation of p about 68% of the time. So if we say that p will be in an interval that extends 0.16 on either side of $X/10$, we will be right at least 68% of the time. When X falls equal to 7, this interval is from 0.54 to 0.86. So we call the interval from 0.54 to 0.86 a 68% *confidence interval* for p.

Textbook expositions of this method keep "confidence" quite distinct from "probability." They emphasize, moreover, that the confidence coefficient of 68% is ultimately a frequentist probability: it is approximately the frequency with which a certain method produces an interval that covers p. Yet the language encourages us to interpret this frequentist probability as an opinion about p. It is the degree to which we can be confident that p is between 0.54 and 0.86. Most users of statistics see little difference between this and a Bayesian degree of belief.

Confidence intervals are only one method in the repertoire of the frequentist statistician. Another important method is statistical testing, especially the use of goodness-of-fit tests for statistical models. We need not describe such tests here; they are discussed in most statistics textbooks. But they too produce frequentist probabilities (significance levels or *P*-values) that are given a belief interpretation at the level of practice (Box 1980).

4. THREE ASPECTS OF PROBABILITY

Probability is a complex idea. Belief, frequency, and support are three of its aspects, and it has other aspects as well. One way to bring together the many aspects of probability is to emphasize the special situation that occurs when we repeatedly perform an experiment for which we know only the long-run frequencies of outcomes. In this special situation, we know the frequencies, and we know nothing else that can help us predict the outcomes. The frequencies therefore determine odds, or prices for tickets on events. These are more than personal prices; they are fair prices, in the sense that they break even in the long run. Because the frequencies are our only evidence, they also determine well-defined numerical degrees of support for events, or degrees to which it is warranted or rational to believe that the events will happen.

The triangle in Figure 2 symbolizes how the ideas of fair price, warranted belief, and knowledge of the long run hold together, both conceptually and historically. Conceptually, we can start at any point in the triangle and reason in the direction of the arrows. Historically, probability began as a theory of fair odds in games of chance, and ideas of probability (which then meant warranted belief) and frequency were only gradually incorporated into the theory.

In the first part of this section, we use the triangle of Figure 2 to gain a clearer understanding of why the different aspects of probability are aspects of a single concept. Then we use the triangle to review the history of the conceptual development of probability. We conclude with some suggestions for recasting the standard interpretations so as to regain the unity represented by the triangle.

Shafer (1991a) develops these themes further. Shafer (1990) describes the conceptual triangle in more detail. Daston (1988) and Shafer (1991b) discuss aspects of the historical development.

4.1 The Conceptual Triangle

Let us consider how we can move around the triangle conceptually, starting with our knowledge of the long run.

The knowledge of the long run that we have in the special situation described by probability theory is quite extensive. We know long-run frequencies for the outcomes of our experiment. We also know about the rate at which these frequencies are likely to converge, and we know betting schemes are futile. We know we cannot accomplish anything by strategies for compounding bets on successive events at the odds given by the long-run frequencies. No such strategy can assure us of a net gain or give us any reasonable expectation of substantially multiplying our initial stake.

This knowledge of the long run already refers to odds for events in individual experiments. These odds are fair because we break even in the long run by betting at them. By compounding bets, we can derive fair odds for events that involve more than one experiment, and we can study how these odds change as the experiments are performed. This is the arrow upward and to the left in Figure 2, the arrow from knowledge of the long run to fair odds on all events.

Any of the three circles in Figure 2 can be taken as a starting point for the mathematical theory of probability.

Once we have fair odds, or fair prices for tickets on events, we can use these odds or prices as degrees of belief. Because the odds are fair odds, not just personal odds, the degrees of belief are warranted degrees of belief, not just personal degrees of belief. From the properties of the fair odds, we can derive rules for these warranted degrees of belief, which we may call probabilities. This is the arrow to the right in Figure 2. The rules we derive for probabilities are similar to Kolmogorov's axioms and definitions, except that they involve probabilities changing as the experiments are performed, not probabilities conditional on arbitrary events.

From the rules for probabilities, we can deduce the knowledge of the long run with which we began. This is the arrow downward and to the left in Figure 2.

Any of the three circles in Figure 2 can be taken as a starting point for the mathematical theory of probability. The theory of algorithmic complexity theory starts with knowledge of the long run. Kolmogorov's axioms start with warranted belief. Similar axioms have been formulated for fair price.

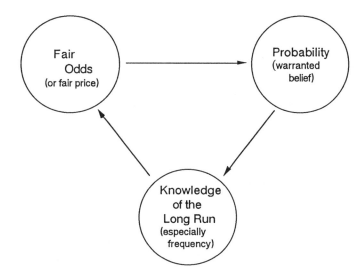

Figure 2: The triangle of probability ideas

The fact that knowledge of the long run, fair price, and warranted belief can each be used as a starting point for the mathematical theory does not mean that any one of these ideas is sufficient for grounding probability in a conceptual sense. The axioms or assumptions we need in order to begin with any one of these starting points can be understood and justified only by reference to the other aspects of the picture. The three aspects of probability are inextricably intertwined.

4.2 The Historical Triangle

The historical development of mathematical probability followed Figure 2, but with fair price as the starting point. The theory began with problems of equity in games of chance, and it only gradually expanded to encompass the ideas of probability and knowledge of the long run.

What we now call mathematical probability began in the 1650s with the work of Blaise Pascal and Pierre Fermat. They were primarily interested in equity—in finding fair odds in games of chance. They did not discuss probability, or the weighing of arguments, which was an important topic at the time. Both probability and frequency were brought into the theory later, by James Bernoulli. In his *Ars conjectandi*, published in 1713, Bernoulli explained that probability is degree of certainty, and he related certainty to equity by saying that an argument has a certain share of certainty as its fair price. He also brought frequency into the theory by proving the law of large numbers.

Bernoulli's moves from equity to degree of certainty and then to frequency are represented by two of the arrows in Figure 2. The third arrow, from frequency back to equity, came much later. Today we are accustomed to saying that the odds given by probability theory are fair because they are odds at which we will break even in the long run. More generally, the expected value $E(X)$ of a random variable X is the fair price of X because it is the price at which we will break even in the long run. This idea appears very late in the probability literature, however. It was first formulated, apparently, by Condorcet in the 1780s (Todhunter 1865), and it did not become popular until the nineteenth century.

The weight of opinion on the foundations of probability theory moved around the triangle even more slowly. Ideas of fairness remained at the foundation of the theory well past 1750. It was only as the probabilistic theory of errors became important in the second half of the eighteenth century that probability, in the sense of rational belief, became fully independent of ideas of equity. An important signpost in this development was Laplace's influential *Théorie analytique des probabilités*, first published in 1812. Laplace interpreted probability as rational degree of belief, and he took the rules for probability to be self-evident. He did not derive them, as his predecessors had, from rules of equity.

In retrospect, Laplace's views look much like the support interpretation, but he did not make the kind of distinction between support, belief, and frequency that we make today. Though he began with the idea of support or rational degree of belief, he did not hesitate to follow Bernoulli in deducing that the long-run frequencies of outcomes will approximate their probabilities.

The frequency interpretation arose in the nineteenth century because of the influence of empiricist philosophy. The empiricists saw fairness, degree of certainty, and rational belief as metaphysical ideas, ideas not grounded in reality. They saw frequency as the only empirical grounding for the theory. So probability should start with frequency. The mathematicians should not pretend, as Bernoulli and Laplace had, to derive facts about frequency from metaphysical ideas about subjective certainty or rational belief.

We also need to acknowledge the subjective aspects of the frequency story. A full account must go beyond the existence of frequencies to the fact that we know these frequencies and nothing more that can help us predict. The randomness of a sequence is not an objective fact about the sequence in itself. It is a fact about the relation between the sequence and the knowledge of a person.

The frequency interpretation became dominant only in the late nineteenth and early twentieth centuries. As its own shortcomings became evident, twentieth-century scholars sought new foundations for the older non-frequentist idea of probability. This produced the belief and support interpretations. The belief interpretation, first advanced by F.P. Ramsey and Bruno de Finetti in the 1920s, went back to the ideas of the pioneers, except that it replaced fair odds and rational degree of belief with personal odds and personal degree of belief. The odds were odds at which a particular person would bet, not odds at which it was fair to bet. This made the interpretation empirically respectable. A person's betting behavior is an empirical fact, not a metaphysical idea like fairness. The support interpretation was less of a departure and more of a continuing defense of Laplace's ideas against the empiricism of the frequentists. John Maynard Keynes and Harold Jeffreys were very influential in this defense.

4.3 Unifying the Standard Interpretations

The historical and practical entanglement of the standard interpretations, together with their unity in the special situation of a sequence of experiments for which we know long-run frequencies, suggests that they should be recast in a way that emphasizes their commonalities.

For the belief interpretation, this would involve returning to probability's original emphasis on *fair* odds rather than *personal* odds. Ramsey and de Finetti's attempt to drop fairness was a mistake. There is no reason for a person to have personal odds at which she would bet on either side. But a person can draw an analogy between her evidence and the special situation where fair odds are known. She can say that her evidence is analogous, in its strength and import, to knowing certain fair odds, which are based on long-run frequencies. This recasting of the belief interpretation pulls it toward both the frequency and support interpretations.

We also need to acknowledge the subjective aspects of the frequency story. A full account must go beyond the existence of frequencies to the fact that we know these frequencies and nothing more that can help us predict. The randomness of a sequence is not an objective fact about the sequence in itself. It is a fact about the relation between the sequence and the knowledge of a person. This point emerges in various ways in the frequentist foundations pioneered by von Mises and Kolmogorov. In Kolmogorov's complexity theory, for example, the complexity of a sequence is defined in terms of the length of the computer program needed to generate it, and this depends on what programming language is used. This means that what is random for a person using one programming language may not be so random for another person. Frequentists tend to minimize this nonobjective aspect of the complexity idea by talking about longer and longer sequences—or even by taking refuge in the idealization of infinite sequences (Uspenskii et al. 1990). But if we refuse to minimize it, we create another point of contact between frequentist and belief foundations.

We can build on this point of contact by emphasizing the ordering of events when we explain the belief interpretation of conditional probability. If we begin with an ordering of events, we have a sequence of events and hence frequencies have a place within the belief interpretation.

With this approach, the three interpretations begin to resemble each other. All three are really about the spe-

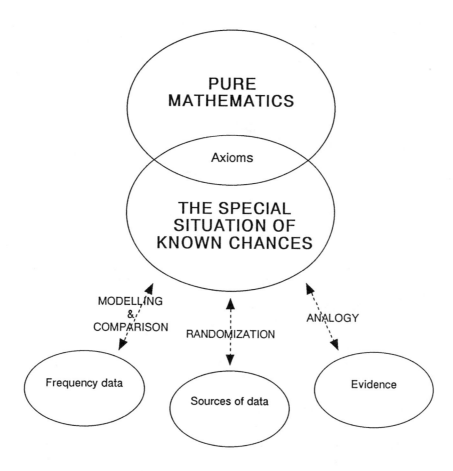

Figure 3: A unified interpretation of probability

cial situation where we have a sequence of experiments with known chances. Most applications of probability, including statistical inference, lie outside this special situation, of course. But we can think of these applications as various ways of relating real problems to the special situation. Much standard statistical modelling amounts to using the special situation as a standard of comparison. Statistical arguments based on sampling or randomization depend on artificially generated random numbers which simulate the special situation. Bayesian analyses are arguments by analogy to the special situation.

Figure 3 summarizes this approach to the meaning of probability. Here probability becomes a story about a special situation involving known long-run frequencies. Various aspects of this story can be made into pure mathematics, and we can base this pure mathematics on Kolmogorov's axioms. The different applications of probability do not, however, depend on different inter-

pretations of the axioms. Instead, they are different ways of using the probability story.

5. LESSONS FOR TEACHING

The main message of this chapter is that probability is a complex idea. It is not simply a set of axioms, nor is it a single interpretation of these axioms. It is a tangle of ideas that took hundreds of years to evolve.

This complexity is evident in textbooks on probability and statistics. A few textbooks manage to take an uncompromising ideological line, either frequentist or Bayesian, but this is hard to sustain. We need to appeal to all the aspects of probability in order to teach the mathematics of probability effectively. We must appeal to frequency in order to explain why probabilities add and in order to explain the significance of the expected value of a random variable. We must appeal to belief when explaining the idea of conditional probability. We must appeal to support when explaining why scientists want to use probabilistic ideas in data analysis.

An understanding of the complexity of probability should encourage humility when teaching the subject. We should be wary of pointing to any particular aspect of probability and saying, "This is what it really means." In particular, we should be wary of telling students that probability is simply a branch of pure mathematics. Probability is not measure theory. It did not begin with Kolmogorov.

The complexity of probability should also make us wary of any strict ideology in teaching statistics. Most elementary textbooks take a sampling-theory viewpoint, but they do not adhere to it strictly, and there are good reasons for this laxness. Some textbooks take a Bayesian approach; here too the teacher needs to be aware that there are good reasons why the proclaimed subjective interpretation is carried only so far.

The main message of this chapter is that probability is a complex idea. It is not simply a set of axioms, nor is it a single interpretation of these axioms. It is a tangle of ideas that took hundreds of years to evolve.

The ways in which probability are used, in statistical inference and elsewhere, are varied, and they are always open to criticism. We should guard, however, against the idea that a correct understanding of probability can tell us which of these applications are correct and which are misguided. It is easy to become a strict frequentist or a strict Bayesian and to denounce the stumbling practical efforts of statisticians of a different persuasion. But our students deserve a fair look at all the applications of probability.

Acknowledgments

Research for this article received partial support from the National Science Foundation through grant IRI8902444 to the University of Kansas.

Glenn Shafer is Ronald G. Harper Distinguished Professor of Business at the University of Kansas. He received an A.B. in mathematics in 1964 and a Ph.D. in statistics in 1973, both from Princeton University. He has taught in the Statistics Department at Princeton and in the Mathematics Department and the Business School at the University of Kansas. He is interested in the history and philosophy of probability and statistics and in the use and limits of probability in artificial intelligence. He is a fellow of the Institute of Mathematical Statistics and a former associate editor of the *Journal of the American Statistical Association*. He has been a Guggenheim fellow as well as a fellow of the Center for Advanced Study in the Behavioral Sciences.

REFERENCES

Barnett, V. (1982), *Comparative Statistical Inference*, second edition, New York: John Wiley.

Box, G. E. P. (1980), "Sampling and Bayes' Inference in Scientific Modelling and Robustness" (with discussion), *Journal of the Royal Statistical Society, Series A*, **143**, 383–430.

Cover, T. M., Gacs, P., and Gray, R. M. (1989), "Kolmogorov's Contributions to Information Theory and Algorithmic Complexity," *The Annals of Probability*, **17**, 840–865.

Cox, R. T. (1961), *The Algebra of Probable Inference*, Baltimore: The Johns Hopkins Press.

Cramér, H. (1946), *Mathematical Methods of Statistics*, Princeton, NJ: Princeton University Press.

Daston, L. (1988), *Classical Probability in the Enlightenment*, Princeton, NJ: Princeton University Press.

Diamond, P., and Rothschild, M. (1978), *Uncertainty in Economics*, New York: Academic Press.

Efron, B. (1978), "Controversies in the Foundations of Statistics," *American Mathematical Monthly*, **85**, 231–246.

Finetti, B. de (1974, 1975), *Theory of Probability*, 2 vols., New York: Wiley.

Jeffreys, H. (1961), *Theory of Probability*, third edition, Oxford: Oxford University Press.

Kyburg, H. E., Jr., and Smokler, H. E., eds. (1980), *Studies in Subjective Probability*, second edition, New York: Robert E. Krieger.

Martin-Löf, P. (1969), "The Literature on von Mises' Kollektivs Revisited," *Theoria*, **35**, 12–37.

Nagel, E. (1939), *Principles of the Theory of Probability* (Volume 1, Number 6 of the *International Encyclopedia of Unified Science*), Chicago: University of Chicago Press.

Porter, T. M. (1986), *The Rise of Statistical Thinking*, 1820-1900, Princeton, NJ: Princeton University Press.

Savage, L. J. (1972), *The Foundations of Statistics*, second edition, New York: Dover.

Schrödinger, E. (1947), "The Foundation of Probability—I," *Proceedings of the Royal Irish Academy, Series A*, **51**, 51–66.

Shafer, G. (1976), *A Mathematical Theory of Evidence*, Princeton, NJ: Princeton University Press.

—— (1985), "Conditional Probability" (with discussion), *International Statistical Review*, **53**, 261–277.

—— (1990), "The Unity of Probability," in *Acting under Uncertainty: Multidisciplinary Conceptions*, ed. G. von Furstenberg, New York: Kluwer, pp. 95–126.

—— (1991a), "Can the Various Meanings of Probability be Reconciled?" To appear in *Methodological and Quantitative Issues in the Analysis of Psychological Data*, second edition, ed. G. Keren and C. Lewis, Hillsdale, NJ: Lawrence Erlbaum.

—— (1991b), "The Early Development of Mathematical Probability." To appear in *Encyclopedia of the History and Philosophy of the Mathematical Sciences*, ed. I. Grattan-Guinness, London: Routledge.

Todhunter, I. (1865), *A History of the Mathematical Theory of Probability*, London: Macmillan.

Tversky, A., and Kahneman, D. (1986), "Rational Choice and the Framing of Decisions," *Journal of Business*, **59**, S251–S278.

Uspenskii, V. A., Semenov, A. L., and Shen, A. Kh. (1990), "Can an Individual Sequence of Zeros and Ones Be Random?" *Russian Mathematical Surveys*, **45**, 121–189.

Walley, P. (1991), *Statistical Reasoning with Imprecise Probabilities*, London: Chapman and Hall.

Chapter 7

The Reasoning of Statistical Inference

Lincoln E. Moses
Stanford University

1. INTRODUCTION

One can think of statistical inference as a collection of methods for learning from experience. The "experience" is typically numerical; the "learning" relates to some population or data-generating process. Statistical inference aims at using the data to augment knowledge about the population or process from which the data come.

Thus, statistical inference "reasons from the data to the population". This contrasts with probability theory, which reasons from knowledge of a population or process, to make statements about samples (i.e., data that might arise).

It is easy to illustrate the two approaches within the framework of n independent Bernoulli trials, each with the same probability of success. Probability problems include:

A1 If $n = 80$ and $P = .7$, what is the probability that at least 50 successes will occur?

A2 How large must n be to give an even chance that $\hat{P}$, the fraction of successes in the sample, will differ from P by at most .1?

Problems of statistical inference include:

B1 From 60 successes observed in 100 trials, can we reasonably conclude that P is not so small as .45?

B2 Knowing from genetics that P is either 3/4 or else 1/2 in this study, which shall we conclude is the truth, if among 100 independent trials we have observed 64 successes?

Reflection on examples B1 and B2 quickly produces the realization that in each one the conclusion adopted must carry some risk of being wrong! This is quite characteristic; typically a statistical inference carries uncertainty with it, and a key part of the inference may be to characterize that uncertainty. Not surprisingly, one often uses probability theory to attack (parts of) problems of statistical inference, though it is not the only useful tool. Simulation can be very helpful. So can clear analytical reasoning.

One key part of statistical inference is not amenable to probability theory, or even to simulations; that key part is the realism of the mathematical formulation employed to represent the real-world phenomena from which the data come. Thus, in problems B1 and B2, the model of independent Bernoulli trials, each with the same (but unknown) probability P of success, may apply to certain engineering contexts, but not to some psychological contexts in which learning or fatigue effects may produce trends in P or dependence between successive trials.

One can think of statistical inference as a collection of methods for learning from experience. The "experience" is typically numerical; the "learning" relates to some population or data-generating process.

The practitioner of statistical inference cannot shirk the question of the realism of the model used, except on pain of failure.

Often a given data-generating set-up can be plausibly mathematized in more than one way, and resulting inferences may differ importantly. Choices among such alternatives are also a part of statistical inference, as we shall see. Reliance on models that are relatively non-committal is often preferred, so as to reduce liability to gross error. In general, the greater the ignorance about the process, the more attractive such noncommittal formulations become.

We explore these and other matters in the pages that follow. We begin by presenting four examples and taking note of issues that they offer. Then we look at some of the most-used stratagems of statistical inference, and we continue with a discussion of various criteria for preferring one method of statistical inference over another in varied circumstances. Then we offer remarks concerning the place of theory in all this, and finally we close with some observations about teaching statistical inference.

2. SOME EXAMPLES

To aid in developing the key ideas and methods of statistical inference, we begin by examining some varied, but typical, examples, after which some general methodological theory and principles will be easier to appreciate.

Example 2.1: Down's Syndrome and Acute Childhood Leukemia

Krivit and Good (1957) had a clinical impression that leukemia occurred surprisingly often among children with Down's syndrome, a congenital condition. At the time they were writing, it was important to know whether this was true; Down's syndrome was known to be genetic in origin, and if leukemia had a strong association with it, that fact would cast some light on the etiology of leukemia. By sending questionnaires to colleagues in the pediatric profession in the United States, they identified 34 cases of leukemia occurring during a 4-year period in children below the age of four who also had Down's syndrome.

Census figures showed that during those years that age group contained about 17,500,000 children on the average. In the age group, about 1 child in 20,000 develops leukemia per year; each year, about 1/20,000th of 17,500,000, or 875, cases of leukemia occurred in this age group; and over the four-year period, about 3500 leukemia cases occurred in children in this age group. A conservatively high estimate for the incidence of Down's syndrome at birth is about 1 in 300, so that *if there were no association between the two conditions*, 1 in 300

of these 3500 leukemia patients should also have Down's syndrome. Thus, on the hypothesis of independence of the two conditions, leukemia and Down's syndrome, the expected number of children showing both conditions would be about $(3500)/(300) = 11\ 2/3$. The Poisson distribution is an appropriate model for the number of occurrences in the four years. The hypothesis of independence ascribes mean 11 2/3 to that Poisson distribution. The actual occurrence of so many as 34 from that distribution would be virtually miraculous, since the probability of so many cases is less than 1×10^{-8}. Thus, the data urge rejecting the model. The authors correctly concluded that some connection between the mechanism causing leukemia in children and the mechanism causing Down's syndrome did in fact exist. Later research revealed more about the congenital connection between these disorders.

We pause to note certain features of this example. The calculation rested on these assumptions:

(i)   The Poisson distribution is applicable to data of this kind—counts of independent occurrences of a rare event.

(ii) Each year about 17.5 million children were in the age group.

(iii) Leukemia incidence is about 1/20,000 per year.

(iv) Incidence of Down's syndrome is not more than one in 300 births.

Further, the calculation introduced and used the hypothesis that the two diseases were statistically independent; that is,

(v) $\text{Prob(both } L \text{ and } D) = \text{Prob}(L) \times \text{Prob}(D)$.

When the data were found in virtual contradiction with the model, the weak link, proposition (v), was regarded as false. Formally, assumptions (i)–(iv) were assumed true (or nearly true), with good reason, and (v), the hypothesis, was exposed to the data. When the data—here $X = 34$—clashed with the hypothesis-including model, that X was an observation from the Poisson distribution with parameter 11 2/3, the hypothesis was rejected as inconsistent with empirical observation. The pattern in which data confront a model, parts of which are taken as true and parts of which are on test, recurs generally

Infant	Waterbed (x)	Bassinet (y)	Difference ($d = y - x$)
1	0.89	1.36	0.47
2	0.77	1.66	0.89
3	0	0.11	0.11
4	0.65	1.44	0.79
5	0.88	1.63	0.75
6	1.36	1.52	0.16
7	1.22	1.53	0.31
8	0.30	0.48	0.18

Table 1: The number of apnic episodes per hour of sleep for eight infants on a bassinet and on a waterbed. Data for Example 2.2.

in areas of statistical inference labeled hypothesis testing or significance testing.

Finally, some attention is due to the data, $X = 34$, in the example. Krivit and Good were at pains to remove duplicated cases, corresponding to children seen by specialists in more than one part of the country, and in this they presumably succeeded. They worked hard to get *all* the cases of acute childhood leukemia combined with Down's syndrome in the USA during the four-year period. But good as their search may have been, they could hardly have succeeded completely. So we must suppose that the number of children below age 4 with both conditions was actually greater than 34. Observe that the conclusion they reached—some association—is not harmed by this possibility, though an *estimate* of the strength of association would be too small.

Example 2.2: Premature Infants, Waterbeds, and Apnea

Premature infants are likely to have interruptions in their breathing while asleep. Such an interruption is called *apnea*. To alert the nursery staff to such episodes, the infant's breathing may be monitored by an automatic device that signals an alarm if a long period of apnea occurs. Reduction of the frequency of apnea episodes has been observed when the premature infant sleeps on a waterbed provided with gentle periodic motion, rather than on an ordinary bassinet mattress. To check out this observation rigorously, Korner et al. (1978) monitored each of eight premature infants for 24 hours, divided into alternate six-hour periods on the waterbed and on the bassinet mattress. Four of the infants started on the waterbed, and four started on the bassinet mattress. The apnea monitor was set to give an alarm if an episode

exceeded 20 seconds. The number of alarms during periods of sleep was counted under the two conditions; then those counts were divided by the number of hours that the infant was asleep. Thus each infant had two scores for apnea alarms per hour of sleep, one for the periods when the infant was on the waterbed and one for the periods when the infant was on the bassinet mattress.

The data appear in Table 1. Inspection of the column of differences shows a consistently greater alarm rate on the bassinet. The average difference $\overline{d}$ is 0.455 alarm per hour of sleep. But is this consistent difference, based on only eight infants, to be regarded as convincing? Might the consistency easily be only an accident of chance?

Several standard methods might be applied to these data:

(1) The *sign test* would lead to the inference that the indicated superiority of the waterbed should not be regarded as a fluke of chance, but should be accepted as evidence of a real effect. The sign test reasons as follows: Each infant had two apnea rates—one for the waterbed, one for the bassinet. Of course, apnea rate varies from one 6-hour period to another, so some difference between the two rates of an infant must be expected. If the two modes were actually equivalent with regard to apnea rate, then for any infant, the probability would be 1/2 that the waterbed rate is the smaller one. For all *eight* to have a lower apnea rate on the waterbed would be very unlikely, an event of probability $(1/2)^8 = 1/256$. That all eight infants would show the bassinet rate as the smaller would also be an event of probability 1/256. Thus, the probability of getting all outcomes to agree in sign is twice 1/256, and we say that the data attain two-sided significance equal to 1/128.

(2) The sign test might be applied in a different sense. If the infants could be regarded as a random sample from a defined population of premature infants, then the sign test would directly infer that the median difference in that population could not reasonably be zero, for if it were the data would consist of eight observations, all of which lie on the same side of the median. Such samples occur on average only two times in 2^8. Rather than believe that we have such a rare sample, we may *reject* the hypothesis that the median is zero; we do so at level of significance 2×2^{-8}, or .008. Further, we would reach this same conclusion for any hypothetical value of the median that was less than .11, for all eight differences would still lie on the same side of such a median; also any hypothetical median larger than .89 would similarly be rejected at level .008. So the interval (.11, .89) is a "confidence interval" for the median of the population of differences from which these eight are a random sample; the confidence attaching to that interval is .992 = 1 − .008.

(3) The *matched-pairs t test* would be exactly applicable if the differences were known to be randomly drawn from some normal distribution with unknown mean μ and unknown variance σ^2. Inferences could relate to testing whether $\mu = 0$, or to calculating a range of values within which one can have high and stated confidence that the value of μ lies.

(4) *Wilcoxon's signed rank test* offers another standard way of analyzing these data, but we will not illustrate the method.

(5) Other approaches, less standard, can be proposed.

The example raises a number of themes of interest. First, there is no general basis for stating which method is "best." If indeed the differences come from some one normal distribution, then the *t* test is best. If they come from some one double-exponential distribution, then the sign test is best. If they come from a logistic distribution, then the signed rank test is best. If the data come from some other distribution, preference among the three methods would require special investigation. In such a situation, one does not ordinarily know the population from which the data come. An important part of statistical theory addresses questions that surround choice among procedures in such ambiguous circumstances. Such investigations often deal with the large-sample case, for as in other mathematical areas, asymptotic analysis can often produce useful approximate results where exact methods are too ponderous to apply.

Example 2.3: Comparing Two Treatments, Applied to Two Different Groups of Subjects

Whitley et al. (1977) reported the results of an experimental therapy for an often fatal, previously untreatable, disease, herpes virus encephalitis. They had allocated 28 patients randomly between the new treatment, Ara-A, and a placebo treatment.* The outcome was as follows:

	Died	Lived	
Ara-A	5	13	18
Placebo	7	3	10
	12	16	28

We see that 13/18, or about 72 percent of the Ara-A patients survived, whereas only 30 percent of the placebo patients did.

Before recommending this therapy for wider use, it is wise to consider whether the encouraging experimental outcome might be only a result of chance, since the samples were not very large.

So we shall test the hypothesis that Ara-A is physiologically inactive—the placebo under a different name. Supposing that hypothesis to be true, we would explain the distribution of the 12 deaths between the two groups as a simple result of the random assignment of the subjects to the two groups: Unknown to anyone, 12 of these patients had the disease so severely that they were destined to die, and 16 did not; the random assignment put 7 of the fatal cases in the placebo group.

Is this explanation reasonable? If so, we could conclude that Ara-A is the placebo under a different name. Or is that explanation so strained that we must reject it and conclude that some causal factor produced the lower death rate in the Ara-A-treated group? To determine how chance might be operating, we begin by showing *all the ways* in which the 12 fatal cases could have been distributed by luck. The 11 panels of Table 2 show all the possible outcomes. Each panel shows 12 deaths distributed among 10 placebo patients and 18

---

*The reader may wonder whether random assignment is suspect here, since the 18 to 10 break is rather unequal. Calculation shows that inequality of that degree or more is to be expected in almost one-fifth of instances where 28 assignments are made by coin-flip, so no alarm is warranted.

(a)

	D	L
A	12	6
P	0	10

(b)

	D	L
A	11	7
P	1	9

(c)

	D	L
A	10	8
P	2	8

(d)

	D	L
A	9	9
P	3	7

(e)

	D	L
A	8	10
P	4	6

(f)

	D	L
A	7	11
P	5	5

(g)

	D	L
A	6	12
P	6	4

(h)

	D	L
A	5	13
P	7	3

(i)

	D	L
A	4	14
P	8	2

(j)

	D	L
A	3	15
P	9	1

(k)

	D	L
A	2	16
P	10	0

Table 2: All possible outcomes entailing twelve deaths among ten placebo and eighteen Ara-A subjects.

Ara-A patients. Panel (k) would provide the strongest indication of effectiveness of the drug, with $2/18 = 11$ percent deaths on Ara-A versus 100 percent deaths on the placebo. The outcome next most favorable to Ara-A is panel (j), then (i), and then (h), which presents the outcome actually observed in the experiment.

To answer the question, "Might an outcome as favorable as (h) (or even more favorable) readily occur by chance?" we must find the probabilities for panels (h), (i), (j), and (k) under *random* assortment of the 12 deaths into the two groups. If the sum of these four probabilities is a small number, like 0.046, we could claim statistical significance and regard the outcome as showing an actual drug effect.

The random assignment of patients to the treatments is central to the ability to rely on the significance test.

Mathematicians long ago solved the combinatorial problem of counting the relative frequencies of such tables and thus finding the level of significance that attaches to any observed result. (That combinatorial solution is known in the statistics literature as Fisher's exact test.) If we did *not* have a mathematical solution, we could estimate the desired probability by simulation: Construct a deck of playing cards consisting of twelve black ones (deaths) and 16 red ones (survivors), shuffle the deck thoroughly, deal out 10 cards (the placebo group), and count the number of black cards among them. The result defines which of panels (a)–(k) has occurred. A few

thousand such shuffles (or computer versions of the same experiment) would provide an estimate of the attained significance level. A third approach to evaluating the probabilities uses asymptotic theory, which produces an approximate evaluation of that significance level. The name of that asymptotic approximation here is "chi-squared."

Several themes are implicit in this example. First, the *random* assignment of patients to the treatments is central to the ability to rely on the significance test. If instead, assignment to Ara-A or placebo were made by clinical judgment, we would be left with the nagging worry that the stronger patients had been more or less systematically assigned to Ara-A, with that fact being the (unrecognized) explanation for the treatment's apparent success. Other ways of assigning patients to the two treatments would leave the question "What scope was there for human judgment to assign the stronger patients to the new therapy?" Random assignment forecloses these doubts.

Second, the example is typical of a very large number of actual applications in laboratory, clinical, engineering and social investigation; the need to compare two success rates arises in many contexts. In some of these contexts, routine experience with an often-used procedure can produce a lot of worthwhile information about control groups. A client of the author's once had data like these from a mouse experiment:

	+	−		
Treated	12	68	80	$\hat{P}_T = .15$
Control	1	24	25	$\hat{P}_C = .04$

The difference between $\hat{P}_T$ and $\hat{P}_C$ does not reach statistical significance at any usual level, chiefly because the control sample (25 observations) is too small to give a sufficiently stable estimate of P_C. But the investigator pointed out that $\hat{P}_C$ had averaged .08 in scores of previous experiments with as much stability as should be seen under random sampling. Moreover, if one compares $\hat{P}_T = .15$ in this experiment with that (nearly) known control rate $P_0 = .08$ (twice as large as the .04 seen in this experiment), then the difference is statistically significant. What should he do? The example raises quite forcefully the relevance of prior information, when it is available, and points to the desirability of incorporating it well. (The advice given was to publish both analyses and tell the reader which one the investigator believed.) Bayesian methods, treated later in this chapter, are well suited to situations of this kind.

Example 2.4: Measuring Blood Volume and Liver Function

Certain organic dyes can be injected into a person's bloodstream, and then detected and measured in blood samples drawn subsequently. Some such dyes are unaffected by any part of the body, except for being trapped and destroyed by the liver as the blood passes through that organ. That action of the liver causes the gradual disappearance of the dye from the blood. Theoretical analysis and practical experiment show that U, the concentration in the blood x minutes after injection, follows the curve

$$U = c_0 10^{-kx}.$$

In this formula, c_0 represents the concentration at the time of injection (as if the dye were all uniformly mixed with the blood, instantaneously), and k represents the rate at which the liver destroys the dye.

If we now set y equal to the common logarithm of U, we have the linear regression function

$$y = \log U = \log c_0 - kx.$$

Actual observations on y will have experimental error; we would represent the data as

$$y_i = \log c_0 - kx_i + e_i \qquad i = 1, ..., n.$$

This describes a linear regression of y on x with slope $-k$ and intercept $\log c_0$. Both of these parameters are biologically interesting. The slope, $-k$, is a measure of liver function, for it indicates how rapidly the dye disappears from the blood. The intercept, $\log c_0$, permits estimation of blood volume. Thus if 5 ml of the dye were initially injected and c_0, the concentration immediately following injection, was estimated as 0.001, then the blood volume is estimated to be 5000 ml.

The job of statistical inference here is to use the observations to construct estimates of k and of $\log c_0$ (which we shall call γ_0 from now on) and to give useful indications of how good these estimates are likely to be.

When we have estimates, call them $\hat{k}$ and $\hat{\gamma}_0$, we can construct for each x_i the fitted value

$$\hat{y}_i = \hat{\gamma}_0 - \hat{k}x_i$$

and the residual $\hat{e}_i$, which is the discrepancy between observation and fit at x_i:

$$\hat{e}_i = y_i - \hat{y}_i$$

These residuals are the key to appraising the quality of the estimates and to judging how the the model fits the data. There are many ways to construct the estimates; choice among them should be determined largely by the probabilistic character of the random components e_i. If those can be taken to be normally distributed with mean zero and some common unknown variance, then there is much to commend using "unweighted least squares" as the estimation method. Other specific distributions of the e_i commend other estimation methods. Another approach aims for "robustness" by estimating k and γ_0 in a way that is relatively insensitive to the particular choice of distribution. Further consideration of this example points to many issues of some generality.

First, checking up on the model specification, an idea we have already met in other examples, arises in two ways here. (i) Is the assertion of *linear* regression actually compatible with the data? This can be looked into by plotting the data and then including a quadratic term in the description of the data:

$$y_i = \gamma_0 - kx_i + \ell x_i^2 + e_i \qquad i = 1, \dots, n.$$

If the fit is significantly improved by adding the quadratic term, we must doubt the adequacy of the straight-line model and conclusions based on it. (ii) Are the assumptions concerning the probabilistic properties of the e_i compatible with the data? Here the observed residuals $\hat{e}_i$ can, especially if n is large, give information about the distribution of the unobserved—and unobservable—errors e_i.

A second issue, of much generality, presents itself if the coupling of x_i to y_i arises, as it often does, through *observing* both x_i and y_i for the ith data point. (In the blood dye example we *chose* x_i, the time of observation, and then *observed* y_i; there was no randomness in x_i. If x is chosen but is hard to set accurately—and so does have some randomness in it—some complication of the previous analysis arises.) Examples abound; each subject might produce a measure of aptitude x and a measure of performance y; or each unit might bear a price x and a measure of quality y, etc. These problems have structure quite different from the dye dilution model. There are now two regressions, one describing the average value of y at varying x, and the other describing the average value of x at varying y. Further, the concept of correlation has meaning in this second context.

A third issue that lurks near our example is *multiple* regression. Often, a variable of interest, say y, is known to vary on average with *several* other variables $u, v, w, \ldots$. Extension of our earlier model to this case brings the following model to the fore:

$$y_i = a + bu_i + cv_i + dw_i + \cdots + e_i$$

or, in more felicitous notation,

$$y_i = b_0 + b_1 x_{1i} + b_2 x_{2i} + \cdots + e_i$$

Tasks of estimation, checking model specification, and the like are again at hand, though complexity is greater.

Clear and significant regression of y on x does not itself imply that if we change x then y will change.

We draw attention to a last issue concerning regression: the temptation to infer cause-and-effect relationships from strong evidence of regression. To say that x causes y means, very roughly, that if we change x, then y will change. But, clear and significant regression of y on x does not itself imply that if we change x then y will change. Consider this example. In adult males both height (y) and girth (y') show marked regression on weight (x). But if we increase x (weight), we will not thereby increase y (height)—though we will increase y' (girth). In general, valid inference about causation from nonexperimental data requires bringing outside knowledge to bear on whatever statistical relationships may be found.

With these examples in hand* we can now turn advantageously to a more systematic presentation of the reasoning of statistical inference.

3. SOME TYPES OF STATISTICAL INFERENCE

A large share of the applications of statistical inference to problems in medicine, engineering, psychology, ecology, agriculture, economics, etc., fall under one of three headings: (i) point estimation, (ii) significance testing (or hypothesis testing), and (iii) interval estimation. It is not unusual for two or more of these approaches to be applicable to the same data set. We take up each of the three kinds of methods in turn, and afterward briefly point to some other kinds of inference, and the situations to which they apply.

3.1 Point Estimation

The idea here is that the observations are assumed to come from a population or a data-generating process that has a certain numerical property of interest (the mean, the median, the 60th percentile, the standard deviation, ...) whose numerical value is unknown. An estimate of that unknown number is desired. The customary language refers to that unknown numerical property as a *parameter*, θ, of the population or process, and the focus of attention is an *estimate* $\hat{\theta}$, based on the observations. The word "point" in point estimation refers to the fact that $\hat{\theta}$, a number derived from the data, corresponds to a point on the real line, or if θ is a vector, denoting $d > 1$ numerical properties, then $\hat{\theta}$, corresponds to a *point* in a space of d dimensions.

Two primary problems are prominent. First, what parameter θ should be estimated? Second, how shall it be estimated from the data?

A good choice of *what* parameter θ to estimate typically rests on the purposes for which the estimate is needed. Describing the distribution of wealth among members of a population by a single number is usually better done with the median than with mean, as the following discussion suggests. Consider two populations of, say, 1500 people identical in wealth distribution except that one of them includes a billionaire. They would have the same median, but substantially different means; the median seems better to reflect the near-identity of wealth distri-

---

*These examples, except for the mouse data, are treated somewhat more fully in my text, *Think and Explain with Statistics*.

bution. On the other hand, from the point of view of the tax collector the mean is a much more germane parameter than the median for comparing the two populations.

Deciding *how* to estimate θ is the second question. It can be easy to propose many estimators, but some will be better than others, in particular circumstances. For example, suppose it is known that the observations $x_1, x_2, \ldots$ are independently distributed, all from some symmetric probability distribution $f_\theta(x)$ with center θ. Possible estimates of θ include:

$$\text{the mean: } \frac{1}{n}\sum x_i = \overline{x}$$
$$\text{the median: } M_x$$
$$\text{the midrange: } \frac{x_{(n)} + x_{(1)}}{2}$$

where $x_{(n)}$ and $x_{(1)}$ are the largest and smallest of the n observations, and

$$\text{the } \alpha\text{-trimmed mean,}$$

which is the average of the observations, excluding the αn smallest and the αn largest.

None of these estimators is best for all symmetric $f_\theta(x)$. Each is best for some such $f_\theta(x)$. Some serve well over a wide variety of distributions $f_\theta(x)$; others over a narrow range.

The examples that we have seen earlier offer instances of point estimation. In the blood volume example with a straight-line relationship, the vector θ comprises γ_0 and k (thus $d = 2$). For the quadratic relationship θ comprises γ_0, k, and ℓ (and $d = 3$). But there is more; the errors e_i are part of the model within which the estimation is undertaken. They can, without loss of generality, be regarded as distributed around a median or mean of zero, for if that center were some other number, a, it could be incorporated into the (unknown) value of γ_0, and the $e_i - a$ would then have center zero.* If the data process was very familiar, it might be justifiable to take as part of the model that the errors had a *known* variance σ_0^2. In that instance the dimensionality of θ would remain 2 for the straight-line case and 3 for the quadratic case. Or, experience might indicate that the variance σ^2, though sensibly constant over the duration of an experiment, varied from one experiment to another; then σ^2

---

*Notice that changing γ_0 by a (the average of measurement errors) does not affect the slope, so the estimate of liver function is not hurt, but the estimate of blood volume would be affected.

also requires estimation, and d, the dimensionality of θ, increases by one, to either 3 or 4.

The mouse experiment in the latter part of Example 2.3 nicely shows the importance of the question "*What should be estimated?*" Is it better to estimate the difference between P_T, the rate in treated mice, and P_C, the rate for contemporaneous control mice, or the difference between P_T and the base rate P_0? This issue can also be approached from another perspective. $\hat{P}_T$ estimates P_T, and both $\hat{P}_0$ and $\hat{P}_C$ estimate the base rate P_0. How best to estimate $P_T - P_0$? Bayesian analysis would judiciously combine $\hat{P}_C$ and $\hat{P}_0$ for comparison to $\hat{P}_T$.

Estimation can also be usefully, if partially, viewed as an algorithmic matter. How shall the observations in the example concerning the liver and the blood concentration of the dye actually be combined to produce estimates of γ_0 and k? The most frequent approach for problems like this one invokes the principle of least squares. In this instance that means, choose those numbers $\hat{\gamma}_0$ and $\hat{k}$ that minimize the sum of squared differences between the observed values $y_1, \ldots, y_n$, and their fitted values, $\hat{y}_i$. That is, choose $\hat{\gamma}_0$ and $\hat{k}$ to minimize

$$Q = \sum [y_i - (\hat{\gamma}_0 - \hat{k}x_i)]^2$$

Another algorithmic approach to the same problem, feasible now with modern computation, would choose estimates of γ_0 and k that minimize

$$A = \text{ median } | y_i - (\gamma_0 - kx_i)|$$

The second set of estimates will not coincide with the first, and will have different statistical properties. Most notably the least-squares estimates will be strongly affected by any extreme values of y in the data set, but the estimates based on minimizing the median of the absolute deviations will not.

This algorithmic perspective on point estimation seems ad hoc, and indeed it is. But the liabilities of adhockery are much assuaged by theoretical investigations of the properties of various estimators under diverse conditions. Such investigations are part of theoretical statistics; they inform the practice of statistical inference and can lead to the abandonment of inferior estimation techniques.

3.2 Significance Testing

The process of significance testing can be described as follows:

(1) The observations are assumed to come from a process for which we have a probabilistic model that we trust.

(2) But the model contains some assertion that we wish to check against the empirical data (the observations). We call that assertion the null hypothesis, H_0.

(3) Some function of the observations (some statistic), T, has a distribution that depends on whether H_0 is true or false.

(4) We study the statistic T; if its observed value is *extremely unlikely* assuming H_0 to be true, we conclude that H_0 is *not* true. (We "reject H_0.") If the value of T is not so extreme as that, we do not reject H_0, supposing that it *might* be true. The two possible conclusions concerning H_0 are "No!" and "Maybe so." The asymmetry results from having resolved to reject H_0 only upon seeing a value of T that is *extremely unlikely* if H_0 is true. So when we reject, we have strong grounds for doing so ("No!"); but when we do not reject, we have only weak grounds for concluding that H_0 is true ("Maybe so").

This insistence on strong evidence before rejecting H_0 aims at ensuring that we will seldom reject H_0 as the result of an accident of chance. We, and other skeptics, can thus have much trust in conclusions that are based on rejecting H_0. The phrase "extremely unlikely" is operationalized by choosing a level of significance, α, typically a small number (perhaps .01 or .05), depending on the context. Then that value of α is used to determine T_α, a possible value of T such that

$$P(T > T_\alpha | H_0 \text{ is true}) \leq \alpha$$

Then if the sample $y_1, \ldots, y_n$ yields a value of the statistic T that does exceed T_α, we declare the results "significant at level α"; that is, we reject H_0 at significance level α.

These ideas appeared in Example 2.1, where rejection occurs for α as small as 1×10^{-8}; in Example 2.2, where rejection occurs for α as small as .008; and in both the herpes virus encephalitis experiment and the mouse experiment mentioned in Example 2.3. In the latter case, choice of the model (or of the statistic) was central to the outcome of the significance test.

The blood dye example (Example 2.4) also evinced the idea of significance testing in the words "if the fit is significantly improved by adding the quadratic term, we must doubt the adequacy of the straight-line model." Within the methodology of least squares it is possible to construct a test at any chosen significance level, α, of the null hypothesis $H_0 : \ell = 0$ (i.e., that the regression has no quadratic component). Notice that deciding this question by a test of significance with small α constitutes a presumption in favor of trusting the straight-line model unless the data give a strong indication of curvature.

Sometimes many significance tests are applied to one body of data, and this poses a new statistical issue, that of many simultaneous comparisons. For example, an epidemiological investigation might ascertain 150 items of information on each of M people with disease D, and also the same items of information on each of N persons who are free of disease D, a comparison group. Now, it is *very probable*, even if disease D has no actual relationship at all with any of the 150 variables studied, that some "statistically significant" differences between the two groups will be found, merely as a result of chance variation. If the significance level is chosen as .05, then $7.5 = .05 \times 150$ such chance results *are to be expected*, on average. This makes it difficult to interpret the results if, say, 10 variables show significant differences at the .05 level. One solution chooses such a small significance level that the chance occurrence of even one significant result among the 150 is only .05. This would call for choosing $\alpha = .05/150 = 1/3000$.

The scope of application for significance testing is wide indeed. Use of the methods involves choices of many kinds, including: the sample size, n; the level of significance, α; the statistic, T; and the way to incorporate related prior information. Once again, the role of theoretical statistics is to cast light on such choices and their consequences in diverse circumstances.

3.3 Interval Estimation

A point estimate of a numerical parameter is useful and natural where a one-number response is required. Thus, in sampling inspection of a batch of items, the contract between supplier and purchaser might vary the price depending upon the percentage of individual items estimated to be defective; then a point estimate is appropriate. Also, in studies that involve many batches of data, each corresponding to one factory, or one school, or one branch office, the need for reducing each batch of data to one or a few numbers gives practical justification to the use of point estimates.

But, typically a point estimate has this shortcoming: The estimate is almost certainly "wrong," in the sense that usually one can be nearly certain that a sample proportion $\hat{P}$ is not equal to the population parameter P, that the sample mean does not equal the population mean, etc. Thus, we are brought to the questions, How close can we suppose $\hat{\theta}$ lies to θ? and How sure can we be of that conclusion?

The method of interval estimation addresses these questions. Let us review the method as we saw it applied in the second treatment of the bassinet-waterbed example. There we postulated a population of differences. Each premature infant who might have been drawn into our sample had such a difference (albeit unmeasured): $d = y - x$, where y and x are the hourly apnea rates on bassinet and waterbed, respectively. This population of values of d necessarily has a median, θ. Using the sign test and our eight observations $d_1, \ldots, d_8$, we constructed an interval of values of θ; it contained all the values of θ that would not be rejected at $\alpha = .008$ by applying the sign test to $d_1, \ldots, d_8$. The validity of this procedure rests upon our sample being drawn at *random* from the population of d's. The t test would give a different confidence interval for θ based on these same d's, and it would depend for its exact correctness on an additional assumption, that the population of values of d from which our sample was randomly drawn had a Gaussian ("normal") distribution. Still other tests would provide other confidence intervals. But always the logic would be the same: The confidence interval, with confidence coefficient $1 - \alpha$, comprises those values of θ that the test would not reject at significance level α on the basis of the data. There is a second aspect to the logic: The confidence coefficient $1 - \alpha$ denotes the "batting average" of the method. That is, in a long sequence of such confidence statements, the percentage of the sample-based intervals that in fact do contain the parameter θ will be $1 - \alpha$. We take comfort in the thought that our particular confidence interval, the one now in hand, has been constructed by so reliable a method, and in this sense we believe that *this* confidence interval contains θ somewhere within it.

Not every hypothesis test produces confidence intervals. Consider the herpes virus encephalitis example. The data there enabled rejecting the null hypothesis that the relative success of the two treatments was equal, but no parameter, θ, entered into consideration. The conclusion was that the Ara-A group had too few deaths in this experiment for chance assortment of lethal cases to be a

reasonable explanation. But without a parameter surely we can't find a confidence interval.

If the 28 subjects *were* a random sample from a population of herpes virus encephalitis cases, then we might postulate two rates: P_T denotes the probability of recovery if treated with Ara-A, and P_C denotes the probability of recovery on the control treatment. And now it would be possible to test $H_0 : P_T = P_C$. Somewhat surprisingly, exactly the same test as before is applicable to testing this hypothesis. It is also surprising that the test cannot be used to give a confidence interval for $P_T - P_C$. (If the samples are quite large, then an approximate confidence interval becomes possible.) But an exact confidence interval for a *different measure* of the contrast between P_T and P_C is constructible. If P is transformed to $\psi(P) = \log\{P/(1 - P)\}$, then the quantity $\psi(P_T) - \psi(P_C)$ has an exact confidence interval. That is, we can choose α and compute two functions of $\hat{P}_T$ and $\hat{P}_C$, say L_α and U_α, such that we can assert with confidence $1 - \alpha$ or more that

$$L_\alpha < \psi(P_T) - \psi(P_C) < U_\alpha.$$

Equivalently, we may give the interval for the "odds ratio,"

$$\exp\{L_\alpha\} < \frac{P_T/(1 - P_T)}{P_C/(1 - P_C)} < \exp\{U_\alpha\}$$

Large values of the odds ratio occur when $P_T > P_C$.

The method of confidence intervals also applies in the regression problem; hypothetical values for k and for γ_0 are readily tested. (These tests are exact if the errors e_i have a common normal distribution, and approximate otherwise.) The tests then can be used to define intervals for k and for γ_0 having desired confidence.

A confidence interval can be used to test a hypothesis in a very natural way. If we have confidence .98 that θ lies between, say, 14 and 19, then that implies that we would reject such hypothetical values as 23 or 12.5 or zero at significance level .02 (or less). The confidence interval is thus more fully descriptive than a mere report that a particular null hypothesis was, or was not, rejected. But, as we have seen, in some problems no confidence intervals are available, though hypothesis testing can be applied.

3.4 Some Other Forms of Statistical Inference

Point estimation, significance testing, and interval estimation may be the most-used modes of statistical inference, but they do not comprise all the methods for reasoning from data. We point briefly to a few other useful modes of statistical inference.

Classification. Considerable pre-training data concerning successful and unsuccessful student pilots may be available. If so, how can it be used to classify a new applicant as likely to be successful, or not? A laboratory test value x tends to be high in persons with disease A and low in persons who do not have disease A. It is natural to set a threshold a, so that if $x > a$, the person is diagnosed as having disease A, and otherwise not. How should that threshold be chosen? On the basis of several morphological measurements $Z_1, Z_2, \ldots, Z_k$ it may be possible (usually) to correctly classify a jawbone as coming from species 1, 2, 3, or 4. How should the data $Z_1, Z_2, \ldots, Z_k$ from already-classified specimens be used to classify a newly discovered jawbone? All of these are examples of classification problems. All use data on some variable(s) from a "learning set" of correctly classified specimens to construct a rule, a function of the variable(s), for classifying future cases. Some of the methods resemble significance testing, but others do not.

Cluster Analysis. Here n specimens each furnish data on many variables, and the effort is to identify subsets of specimens that are similar to each other; such subsets are called "clusters." Problems of this kind also go under the name pattern recognition. This problem and classification usually involve data comprising many variables for each subject, so-called multivariate data. The next problem type also has this character.

To mistakenly treat description as "routine" is almost surely to botch the job.

Finding More Parsimonious Representations of Multivariate Data. An educational psychologist studying high school seniors might obtain data on a large number of characteristics: grades in various courses, scores on various aptitude tests, attitudinal scales, questionnaires about recreational activities, work activities, and friendships, etc., possibly 50 variables for each student. Can some few composite summaries of these variables be found, so that the information in the 50 variables is (nearly) fully captured in those fewer composite scores? A little reflection suggests that this may be a reasonable hope. Once we have seen a student's grades in several courses we have a pretty good idea of that student's academic work, and information about grades in

several additional courses is likely to be redundant. More generally, it is hard to think of 50 distinct attributes of a high school student that would *not* involve considerable redundancy. Some methods for dimension reduction are factor analysis, principal components analysis, multi-dimensional scaling, and canonical correlation analysis.

A Word About Description. Each of the last three inference methods has had a strong flavor of *description*. Sometimes "descriptive statistics" and "inferential statistics" are contrasted, with an implication of description as the lesser of the two. That attitude is mistaken. Good statistical description is demanding and challenging work; it requires sound conceptualization, and demands insightfully organizing the data, and effectively communicating the results; not one of those tasks is easy. To mistakenly treat description as "routine" is almost surely to botch the job.

4. WHICH MODEL? WHICH ANALYSIS?

Repeatedly we have seen how a given data set can be viewed from more than one perspective, can be represented by a model in more than one way. Quite commonly no unique model stands out as "true" or correct; justifying so strong a conclusion might require a depth of knowledge that is simply lacking. So it is not unusual for a given data set to be analyzed in several apparently reasonable ways. If conclusions are qualitatively concordant, that is regarded as grounds for placing additional trust in them. But more often, only a single model is applied, and the data are analyzed in accordance with it. Thus, it seems useful to give attention to strengths or weaknesses of various kinds of models.

4.1 Some Desirable Features of Models

Desirable features in a model include (i) tractability, (ii) parsimony, and (iii) realism. That there is some tension among these is not surprising.

Tractability. A model that is easy to understand and to explain is tractable in one sense. Computational tractability can also be an advantage, though with cheap computing available not too much weight should be given to it.

Parsimony. Simplicity, like tractability, has a direct appeal, not wisely ignored—but not wisely over-valued either. If several models are plausible and more than one of them fits adequately with the data, then in choosing among those, *one* criterion is to prefer a model that is simpler than the other models.

Realism. This notion has two rather different aspects, both important. First, does the model reflect well the actual data-generating process? This question is really a host of questions, some about the distributions of the random errors, others about the mathematical relations among the parameters. The second aspect of realism is sometimes called robustness: If the model is *false* in certain respects, how badly does that affect estimates, significance test results, etc., that are based on the flawed model?

Another aspect of realism that is often overlooked arose in the mouse experiment. The particular data or estimate or significance test at hand may really be one of many related problems; it is then unwise to regard it as an isolated integral problem-in-itself.

4.2 Some Broad Principles of Statistical Analysis

We have already met the idea that a given body of data can be modeled and analyzed in more than one way. But even a fixed given model may permit alternative analyses, and choice among them may deserve thoughtful consideration.

If the model provides that $x_1, x_2, \ldots, x_n$ are independent observations from some symmetric distribution (say, a normal distribution with mean μ and standard deviation σ), then, as we have already seen, we might estimate the center in many ways. Similarly, different estimators might be used to estimate the unknown standard deviation. In the bassinet example we pointed to various ways of testing that the center of the distribution of differences was zero. How can wise choices be made among the many possible strategies for data analysis?

A considerable literature in mathematical statistics treats questions of this kind. We cannot summarize it adequately here, but we offer some salient ideas.

It Is Desirable to Take Full Account of the Data. The principle is more easily stated than applied, because "full account" is problem-specific. Thus, the mean $\bar{x}$ (and not the median, M_x) takes "full account" of the sample information about μ if $x_1, \ldots, x_n$ are independent observations from a normal distribution, but it is exactly the other way around if the observations come instead from a double-exponential distribution. Despite such particularity, useful ideas follow from the princi-

ple. Thus, it immediately raises doubts about practices like throwing away data to make sample sizes equal, or omitting from the analysis of a large number of related data sets all those that have fewer than 5 observations, or fewer than 10. Such rules may come into being for good reasons—like reluctance to treat large and small samples alike; but the correct response is more likely to call for properly taking account of their different sizes, rather than omitting them entirely.

Resistance (to maverick data) Is Often Desirable. Here the word "resistance" has a technical meaning. An estimate $\tilde{\theta}$ based on observations $x_1, \ldots, x_n$ is said to be *resistant* if changing one (or a few) of the observations by a very large amount does *not* exert a large effect on $\tilde{\theta}$. The median M_x is an example. If we have a sample of, say, 17 observations, all different, then M_x is the observation that is ninth when they are arranged in increasing order: $x_{(1)} < x_{(2)} < \cdots < x_{(9)} < \cdots < x_{(17)}$. The sample contains eight larger observations, on its right, and eight smaller ones on its left. Now let any one observation on its right be grossly increased, to (say) a thousand times the largest value in the sample. The observation in the middle is still in the middle, and so the median is not affected at all! If instead, the observation that was so grossly increased came from the left of $x_{(9)}$, we would now have seven on the left of $x_{(9)}$ and nine on its right, so the median would become the first observation on the right of $x_{(9)}$; thus the great change in one observation produces a small effect, or none. Thus we speak of the median as resistant.

On the other hand the correlation coefficient is not resistant. If some one data point (x_i, y_i), with neither coordinate being zero, is distorted to (Kx_i, Ly_i) with $|K|$ and $|L|$ very large (possibly through computer error), the sample correlation coefficient is dominated by that single observation. Indeed, as $|KL| \to \infty$, the absolute value of the correlation coefficient tends to 1.0, whatever be the other observations in the sample.

We have used the word "often" in stating the desirability of resistant estimators, because again advantages and disadvantages tend to be situation-specific.

Robustness Can Be Helpful. Again we have a technical word. An estimate or a test is said to be *robust* if its properties are not much distorted when the model underlying the data is altered. The idea can be seen through an example. Suppose we have many observations $x_1, x_2, \ldots, x_n$ from a distribution that is assumed to be normal with unknown mean μ and unknown standard deviation σ, and suppose it is desired to estimate the 90th percentile of that distribution. Two possible estimates are:

$$\bar{x} + 1.28s$$

and

$$x^{(.90)}.$$

In the above $\bar{x} = (1/n)\Sigma x_i$ is the sample mean, s is the standard deviation, where $s^2 = \Sigma(x_i - \bar{x})^2/(n-1)$, and $x^{(.90)}$ is a number that is exceeded by one-tenth of the observations $x_1, \ldots, x_n$. Both estimates work well in large samples if the data actually do come from a normal distribution; but if they come from some other distribution, the estimate based on $\bar{x}$ and s can be seriously in error, whereas $x^{(.90)}$ will continue to be a good estimator of the 90th percentile of the distribution of x. We refer to this latter estimator as *robust*.

4.3 Bayesian Methods

Recall the investigator with the mouse experiment who found $\hat{P}_T = .15$ ($n_T = 80$) and $\hat{P}_C = .04$ ($n_C = 25$). These sample proportions did not differ significantly, though $\hat{P}_T = .15$ *did* differ significantly from a well-determined prior rate, $P_0 = .08$. The example raised forcefully the importance of taking prior information into account. Suppose that the prior rate arose from 32 positive outcomes occurring in a total (in several experiments) of 400 control mice. Then it would be natural to reconsider the data in hand, as comprising 80 treated mice, including 12 positives and 425 (400 + 25) control mice, including 33 (32 + 1) positives. Then the proportions 12/80 and 33/425 could be compared by the usual method, spoken of in Example 2.3. (The difference is significant at $P < .05$.) Many statisticians would regard this analysis as appropriate—more appropriate than restricting the analysis to the 105 animals in the one experiment.

Suppose now that the prior information was somehow less well-documented, and so vague. Suppose the investigator is "quite sure" from long experience that in this strain of mice, positive results in untreated animals are "very unusual." Surely this information too is a relevant supplement to the experimental data of one positive among 25 control mice. But how shall it be taken into account? One approach, reflecting the Bayesian viewpoint, is to get the terms "quite sure" and "very unusual" rendered in numerical form. Perhaps the investigator opines that about .08 is the typical rate in untreated animals, and he bases this judgment on experience reaching back

over, say, 400 untreated animals. It is possible to proceed formally just as before, but things are on a different footing now. The statistics to be compared are a frank mixture of data and opinion (or judgment).

Typically a Bayesian analysis proceeds in much this way. In addition to the data and a probability model for those data (involving some unknown parameter(s), which we call θ), there is also a prior probability distribution for θ, admitted as a part of the model for the analysis of the problem. The frequentist treats θ as an unknown *constant*; the Bayesian regards it as a random variable with a subjective probability distribution. The output of the analysis is a revised prior probability distribution for θ; it is called the posterior probability distribution. Sometimes particular features of that posterior distribution will capture special attention—its mean, or its 99th percentile, or its first percentile, or the interval between those two percentiles, within which interval there is (subjective) probability .98 that θ actually lies.

The role of Bayesian statistics is still evolving. Many statisticians draw on these ideas in some problems, but not in others. One is reminded of the physicist who treated light as a wave on Mondays, Wednesdays, and Fridays, and as a particle on Tuesdays, Thursdays, and Saturdays. (And on Sundays he prayed.)

Clearly this kind of approach to statistical reasoning puts matters on a new footing, mixing opinion (or subjective probability) with observed data. This mixture will be positively welcomed by a strong believer in the Bayesian approach. She may argue that probability is in its true essence a matter of strength of belief or opinion, that frequency data are merely simple and persuasive inputs to the formation and revision of such opinion, and that the proper task is always to give numerical quantification of opinion (or belief, or subjective probability). The strong skeptic is likely to voice misgivings that, when scientific results are an amalgam of testable, reproducible experimental data on the one hand, and opinion on the other, the objectivity of scientific work is in jeopardy.

This writer does not know of any simple formulation that can satisfy all disputants, nor whether in principle one might exist.

In business and military affairs Bayesian methods may seem more natural; acting on guesses, hunches, and opinions, together with data, have long been the necessary practice in these domains. Reasonable quantification of the subjective ingredients—and outputs—could well come as a boon.

The most ardent Bayesian advocate will grant that sometimes there is scant prior information at hand and may therefore choose an "uninformative" or "gentle" prior distribution. If a binomial parameter P is at issue, a gentle prior might be a uniform distribution for P on the interval (0,1). This leads to a posterior distribution for P, after observing r positive outcomes in n independent trials, which has for its mean $\hat{P} = (r + 1)/(n + 2)$, differing somewhat from $\hat{P} = r/n$, which is the frequentist's classical estimate of P for the same data. It is not hard to accept this Bayesian estimator as reasonable, especially when $r = 0$ or n, for it can be uncomfortable to estimate P as 0 (indicating impossibility) or as 1 (indicating certainty); the estimates $1/(n+2)$ and $1-1/(n+2)$ avoid being so extreme. In large samples the Bayesian and frequentist estimates here (and more generally) differ only imperceptibly. The role of Bayesian statistics is still evolving. Many statisticians draw on these ideas in some problems, but not in others. One is reminded of the physicist who treated light as a wave on Mondays, Wednesdays, and Fridays, and as a particle on Tuesdays, Thursdays, and Saturdays. (And on Sundays he prayed.)*

5. THEORETICAL STATISTICS

To this point we have offered examples of statistical inferences, pointed to some major types of statistical inference, and commented on the importance of well-chosen models and analytic tools. But at times, an air of ambiguity has attended the discussion. We have said little about statistical theory; yet the books are full of theory. It is time now to turn to a brief consideration of the place of theoretical statistics in the practice of statistical inference.

By and large the better and fuller the data analyst's understanding of statistical theory, the better can be that person's applied work. But the sophisticated practitioner will draw on theoretical considerations more as a group of landmarks than as a boulevard or highway giving easy passage through the rough, varied terrain of real-world problems. To develop these ideas, we turn to

*This comparison is usually attributed to Jerome Cornfield.

the most belabored (and fruitful) theoretical model.

It is assumed that $x_1, x_2, \ldots, x_n$ are independently and identically distributed with probability density

$$f(x; \mu, \sigma) = (2\pi\sigma^2)^{-1/2} \exp\{-(x - \mu)^2/2\sigma^2\}.$$

This density function specifies $N(\mu, \sigma)$, the normal (or Gaussian) distribution with center μ and standard deviation σ. A typical first course in theoretical statistics proves the following propositions concerning $\bar{x} = (1/n)\Sigma x_i$ and $s^2 = \Sigma(x_i - \bar{x})^2/(n-1)$:

(i) $\bar{x}$ has distribution $N(\mu, \sigma/\sqrt{n})$.

(ii) $(n - 1)s^2/\sigma^2$ has the chi-squared distribution with $n - 1$ degrees of freedom.

(iii) For large n, s^2 is nearly $N(\sigma^2, \sigma^2\sqrt{2/(n-1)})$.

(iv) $(\bar{x} - \mu)\sqrt{n}/s$ has the student's t distribution with $n - 1$ degrees of freedom.

(v) If s_1^2 and s_2^2 come from two such samples (independent) with sample sizes n_1 and n_2, then s_1^2/s_2^2 has the F distribution with $n_1 - 1$ and $n_2 - 1$ degrees of freedom.

These propositions then form the basis for significance tests and confidence intervals relating to μ, σ^2, and σ_1^2/σ_2^2 for normally distributed data. Formally, each of these conclusions is a consequence of the assumed form of the distribution—$N(\mu, \sigma)$—for the observations. Now, in the real world, when the distribution is *not* normal, but instead merely has all its probability on a set contained in a bounded interval, some of these propositions—(i) and (iv)—remain approximately true, for large n, and the others do not. We might summarize this situation by saying that, for bounded random variables when n is large, the distributional properties of $\bar{x}$ for (normal data) are robust, whereas the Gaussian distributional properties of s^2 and s are not. So wise use of the normal theory in wider circumstances requires that we know more theory. (A particular issue demanding *more* clarity concerns the meaning of n "large." Both theory and simulation can help with this.)

The device of assuming that the observations have a distribution from a particular class (like the normal distributions) has advantages and disadvantages. Among the advantages are access to well-understood statistical methods, including maximum-likelihood estimation and likelihood-ratio tests; properties of these standard tools are typically quite desirable when they are validly applicable, and then tests and estimates within this framework often are best possible in one sense or another.

But there are disadvantages. If the specification is trusted unreservedly, then questions of resistance and robustness have been defined out of consideration; maverick data are not admitted in the specification (so resistance is assumed irrelevant); the *given* model is trusted (so robustness questions are assumed not to be relevant). Perhaps the fundamental disadvantage is that it may not be feasible to check the adequacy of an assumed model; in fact, that is more likely than not, in the writer's experience. Choosing methods of analysis for their optimal properties, under assumptions that are not checkable is itself an ambiguous enterprise. Theoretical considerations can often help in sorting out what is sensible and what is not, but rarely do they realistically determine what is *best* in a given situation.

We might summarize by saying that the sophisticated data analyst, concerned with the logic of statistical inference, will draw on theory for what help it can give, while keeping one eye fixed on the data and the other on the model for the data.

6. REMARKS ABOUT TEACHING STATISTICAL INFERENCE

The nature of the subject commends that it be taught with both theory and data continually present and interwoven. This principle should affect selection of text, problems given to students, and material dealt with in the classroom. Real data are ordinarily greatly preferable to synthetic data and artificial examples.

The teaching of mathematics often draws on figures and diagrams, and for good reasons. Similarly, resort to graphical aids can usually benefit statistics instruction, though it can present challenges. First, there ordinarily are two or more ways to plot a data set or a function related to the analysis. And sometimes a particular one of the possible ways will be the best; but it must be sought. Second, care ordinarily needs to be given to mundane questions like the scales of the axes and placement of the origin, questions that can largely be skirted in many mathematical contexts, but which if ignored in statistical contexts may lead to poor visual representation.

To build statistical intuition and to render the subject matter less abstract, less formal, and more directly apprehended by the student, recourse to in-class sampling experiments can be invaluable. The teacher who gives hard thought to the matter can reap great rewards in the form of bringing every student along with concepts like the sampling distribution of a statistic, the power of a test, and the correspondence between a confidence interval and the significance test that generates it. One well-established text (Dixon and Massey 1969) includes suggested class sampling exercises for topic after topic, chapter after chapter. More recently, the entrance of the computer into instruction has enlarged the range of options. Both individual and group efforts can be conceived, where a pseudo-random number generator replaces a private random number table for each student, and the class as a whole can see a theorem, or other construct, physically realized.

Acknowledgments

The author thanks D. L. Bentley, D A. Bloch, B. W. Brown, Jr., J. Q. Denton, F. C. James, F. Mosteller, and R. A. Olshen for their helpful comments on early drafts.

Lincoln Moses is Professor of Statistics and of Biostatistics at Stanford University. He received an A.B. in social sciences in 1941 and a Ph.D. in statistics in 1950, both from Stanford. After two years as Assistant Professor of Education at Teachers College, Columbia University, he returned to Stanford in 1952 to an appointment equally divided between the Statistics Department and the School of Medicine. Professor Moses served as a presidential appointee for two and one-half years as the first head of the Energy Information Administration, the statistics arm of the U.S. Department of Energy. His professional interests include applications of statistics to medical and behavioral science, public policy, and peace studies. He is a fellow of the American Statistical Association, the Institute of Mathematical Statistics, and the American Association for the Advancement of Science, and is a member of the American Academy of Arts and Sciences, the Institute of Medicine, and the International Statistical Institute.

REFERENCES

Cleveland, W. S. (1985), *The Elements of Graphing Data*, Monterey, CA: Wadsworth.

Dixon, W. J. and Massey, F. J. (1969), *Introduction to Statistical Analysis*, third edition, New York: McGraw-Hill.

Gnanadesikan, R. (1977), *Methods for Statistical Data Analysis of Multivariate Observations*, New York: John Wiley.

Gordon, A. D. (1981), *Classification*, London, England: Chapman and Hall.

Hartigan, J. A. (1975), *Clustering Algorithms*, New York: John Wiley.

Hoaglin, D. C., Mosteller, F., and Tukey, J.W. (Eds.) (1983), *Understanding Robust and Exploratory Data Analysis*, New York: John Wiley.

Korner, A. F., Guilleminault C., Van den Hoed, J., et al. (1978), "Reduction of Sleep Apnea and Bradycardia in Preterm Infants on Oscillating Waterbeds: A Controlled Polygraphic Study," *Pediatrics*, **61**, 528–533.

Krivit, W. and Good, R. A. (1957), "Simultaneous Occurrence of Mongolism and Leukemia; Report of a Nationwide Survey," *American Medical Association of Journal of Diseases of Children*, **94**, 289–93.

Lindley, D. V. (1965), *Introduction to Probability and Statistics from a Bayesian Viewpoint, Part 2: Inference*, New York: Cambridge University Press.

Moses, L. E. (1986), *Think and Explain with Statistics*, Reading, MA: Addison-Wesley.

Mosteller, F. and Tukey, J. W. (1968), "Data Analysis, Including Statistics," in *Handbook of Social Psychology*, eds. G. Lindzey and E. Aronson, second edition, Volume 2, Chapter 10, 80–203. Reading, MA: Addision-Wesley.

Seber, G. A. F. (1984), *Multivariate Observations*, New York: John Wiley.

Tukey, J. W. (1977), *Exploratory Data Analysis*, Reading, MA: Addison-Wesley.

Whitley, R. J., Soong, S.-J., Dolin, R., Galasso, G. J., Chi'en, L. T., Alford, C. A., and the National Institute of Allergy and Infectious Diseases Collaborative Antiviral Study Group (1977), "Adenine Arabinoside Therapy of Biopsy-Proved Herpes Simplex Encephalitis," *New England Journal of Medicine*, **297**, 289–294.

Chapter 8

Diagnostics

David C. Hoaglin
Harvard University

1. AIMS OF DIAGNOSIS

Under the broad heading "diagnostics" a variety of approaches enable statisticians to discover whether a tentative summary or analysis is compatible with the data. A leading example arises in regression, where an individual observation can sometimes have disproportionate impact. Because the impact can take many forms, regression diagnostics include a rich array of techniques for detecting such data.

Sections 2 and 3 of this chapter discuss selected basic diagnostic techniques in simple and then multiple regression. We begin with simple linear regression because its structure presents few complications. Indeed, for such (x, y) data the customary scatterplot reveals everything, so that one might expect to need no specialized diagnostics. Still, by focusing on specific aspects of the regression, the diagnostic measures express the impact of each observation quantitatively. Then our ability to compare displays of the numerical measures against the scatterplot allows us to develop a better understanding of what each diagnostic measure detects.

This background helps us when we tackle multiple regression, where the greater dimensionality of the data often prevents scatterplots from revealing important structure. The diagnostic techniques in our basic toolkit provide a good start in coping with the complexities of multiple regression that occur in practical applications.

Many of the techniques of regression diagnostics arise from a few basic concepts:

- *leverage* — the extent to which an observation's given value of y determines the corresponding predicted value;

- *residuals* — the difference $y_i - \hat{y}_i$ between the observed value (y_i) and the fitted value ($\hat{y}_i$) at the individual data points;

- *deletion* — the process of leaving out each observation in turn and comparing the resulting n regressions, each based on $n - 1$ observations;

- *partial regression* — in multiple regression, treating each explanatory variable in turn as if it were the last variable to enter the regression equation;

- *constructed variable* — an additional explanatory variable, created specifically to reveal a particular type of structure in the residuals (either via a scatterplot or a simple regression).

This chapter has space to offer only a few examples of the application of these fruitful ideas.

Fitting a least-squares regression line to a set of data can give a rude surprise when the data contain one or more highly deviant observations. Diagnostic measures express the impact of each observation quantitatively.

Broadly, both in regression and in other methodologies, diagnostic techniques allow the data to criticize or falsify the tentative statistical model. Besides showing whether individual observations have undue impact, they also reveal departures from

- linearity (e.g., y increases as x^2, rather than as x),

- constant variability throughout the data (e.g., the precision of measurement may be roughly a constant percentage of the quantity being measured),

- additivity of effect (e.g., the contribution of one ingredient enhances the contribution of another),

- a customary distribution (e.g., the data have heavier tails than a normal distribution), and

- other assumptions embodied in a model.

Suitable plots of residuals provide ways of probing for nonlinearity and nonconstant variability. In the setting of a two-way layout, one constructed variable summarizes a common form of nonadditivity. Section 4 discusses this technique, along with probability plotting — to check whether data or residuals seem to depart from an anticipated distribution. After careful use of such diagnostics, followed by any indicated revisions in the model, one can proceed to the stage of making (formal) inferences with some assurance that the model and the data are not working at cross-purposes.

The regression diagnostics that we discuss gain much of their power from the ease with which they can be implemented in least-squares linear regression. Thus they have come to play a special role in overcoming some of the inherent defects of classical regression. For some types of analyses, we have the alternative of resistant or robust techniques of fitting, which are less sensitive to anomalous observations and departures from assumptions. Chapter 9 focuses on such techniques, so we discuss them only briefly in Section 5.

In certain situations, chiefly not regression, concentrating on sensitivity to individual observations does not yield very informative diagnoses. Instead, it may be possible to calculate an estimate or other measure from each of several diverse subsets of the data. This approach underlies the "principle of cooperative diversity," which Section 6 briefly describes and illustrates with a simple example.

2. LEVERAGE AND INFLUENCE IN SIMPLE REGRESSION

Fitting a least-squares regression line to a set of (x, y) data can give a rude surprise when the data contain one or more highly deviant observations. Even when we have a moderate amount of data, a single data point can cause a lot of trouble. The regression line may have a slope or

an intercept very different from what most of the data indicate. We illustrate this phenomenon with an example having one data point that has much greater impact than the others in determining the fitted line.

For each of 12 large federal agencies, Table 1 gives the agency's estimate of the number of full-time employees involved in public affairs activities in 1976 and the associated cost (in dollars). Public affairs activities include providing press information and preparing advertising, exhibits, films, publications, and speeches for release to the public. After allowing for some fixed costs, we might expect the cost per employee to be roughly the same in the various agencies, and so we treat cost as the response variable (y) and the number of employees as the explanatory variable (x).

In Figure 1 the scatterplot of the data immediately reveals that the unusually extreme x-value of the observation (x_1, y_1) for the Department of Defense will allow a large change in the observed y-value to produce a large change in the regression line. When, in Figure 2, we drag y_1 down from its observed position (•) to a much lower position (∘), the regression line flattens dramatically. The line no longer follows the general slope suggested by the other data points. An equal change in any one of the other y-values would have a much less pronounced effect. As the data stand (in Figure 1), however, the point for the Department of Defense seems compatible with a straight line based on most of the other points. By contrast, the cost for the Department of Health, Education, and Welfare seems very high for the number of employees. Also the point for Congress seems somewhat low, perhaps because that cost figure does not include all operating expenses.

In general, whenever an observed y-value is clearly inconsistent with the apparent relationship between y and x, we want to call attention to the anomaly. Further investigation of the circumstances surrounding such a data point could uncover a number of possible explanations, such as:

- The y-value (or the x-value) was recorded incorrectly.

- Someone made an error entering the data into the computer.

- Some unusual event or situation made that data point unlike the others.

- That data point is entirely correct, and a straight

Agency	i	Number of Employees (x)	Cost (y) in Dollars
Defense	1	1486	24,508,000
HEW	2	388	21,000,000
Agriculture	3	650	11,467,300
Treasury	4	202	5,798,235
Congress	5	446	5,663,174
Commerce	6	164	5,683,609
Energy Research and Development Admin.	7	128	5,236,800
NASA	8	208	4,500,000
Transportation	9	117	2,913,509
HUD	10	69	2,455,000
White House	11	85	2,300,000
Veterans Admin.	12	47	1,313,300
Totals		3990	92,838,927

Table 1: Public affairs activities in 12 U.S. Government agencies, 1976. For most agencies the cost includes both salaries and operating expenses. Available records do not indicate total operating expenses for employees of Congress involved in promotional activities. From *The Boston Evening Globe*, April 27, 1976, p. 2.

line is not an appropriate description of the relationship between y and x over the given range of x-values.

Although a simple scatterplot of y against x readily reveals the presence of such anomalous data points, we usually gain by having a systematic way of detecting them. For one thing, a plot may not automatically accompany every regression, especially ones that are supposed to be routine, even when having the data in a computer makes plotting easy. More importantly, when we use multiple regression to investigate the simultaneous contribution of three or more explanatory variables, we have no guarantee that the usual plots will reveal anomalous observations, because such points may not stand out on any pair of coordinates.

A variety of regression diagnostics help users of regression analysis better understand data and models. Because an appreciation of diagnostics can greatly enrich applications, this section explores a core of techniques that should be accessible to beginning users of regression. The data on Public Affairs Activities provide the framework for introducing leverage and the hat matrix in simple regression. After discussing basic aspects of residuals, we consider the notion of leaving out each observation individually, and then we apply this device to study influence on fitted values and to define studentized

residuals. Section 3 extends these ideas to multiple regression.

An Elementary Account of Leverage

As we saw in Figure 2, some observations may have a large impact on the simple regression line

$$\hat{y} = b_1 + b_2 x = \bar{y} + b_2(x - \bar{x}). \quad (1)$$

Here the data are $(x_1, y_1), (x_2, y_2), \ldots, (x_n, y_n)$, and

$$\bar{x} = \frac{1}{n} \sum_{i=1}^{n} x_i,$$

$$\bar{y} = \frac{1}{n} \sum_{i=1}^{n} y_i.$$

If we express cost in dollars, the solid line has slope 15,566 (a plausible per-person salary figure in 1976), whereas the dashed line has slope 2811, and yet we changed only y_1 (decreasing it by \$20,000,000). The ease with which we twisted the fitted line demonstrates that least-squares regression provides no protection against such quirks of the data as deviant x-values. Plotting the data, however, as in Figure 1, enables us to see both expected and unexpected features, and so we would notice that the point for Defense stands apart from the rest. We almost always benefit from plotting the data.

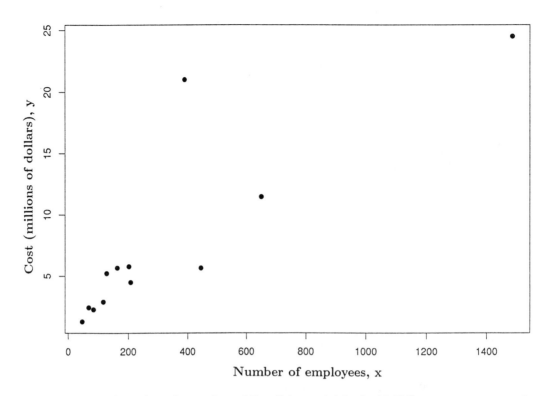

Figure 1: Cost versus number of employees for public affairs activities in 12 U.S. government agencies, 1976.

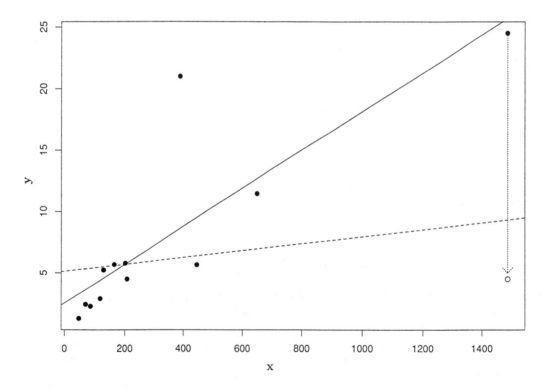

Figure 2: The dashed line is the regression line corresponding to the new position of the point ∘.

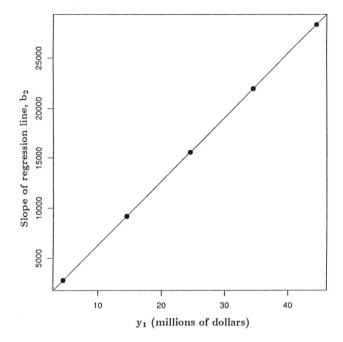

Figure 3: Relation of the slope of the regression line (b_2) to the value of y_1.

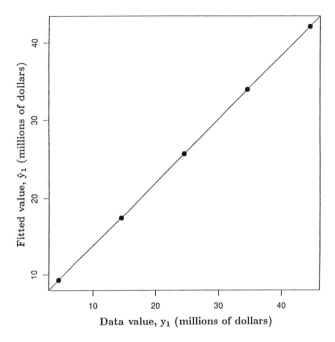

Figure 4: Relation of the fitted value $\hat{y}_1$ to the data value y_1. The slope of the line is 0.819

For a more systematic look at how changing y_1 affects the slope, Figure 3 shows b_2 as a function of y_1 (holding the rest of the data fixed). The five points plotted as circles correspond to $y_1 = 24,508,000 + k(10,000,000)$ for $k = -2, -1, 0, +1, +2$. The line summarizes the relationship and illustrates a basic fact of least squares: *fitted regression coefficients are linear functions of the y-values.* Similarly, in Figure 4, the fitted value $\hat{y}_1$ is a linear function of the observed value y_1 when the other y-values remain fixed.

The relation of $\hat{y}_i$ to y_i, as in Figure 4, leads us easily to a quantitative measure of potential impact known as leverage. The idea is to assess an observation's ability, in a more or less mechanical sense, to pull its fitted y-value toward its observed y-value. We first express the regression line in the more convenient centered form and focus on fitted values:

$$\hat{y}_i = \bar{y} + b_2(x_i - \bar{x}). \tag{2}$$

For b_2 we now substitute a standard expression for the least-squares slope,

$$b_2 = \frac{\sum_{j=1}^{n}(x_j - \bar{x})y_j}{\sum_{k=1}^{n}(x_k - \bar{x})^2}, \tag{3}$$

to obtain

$$\hat{y}_i = \bar{y} + (x_i - \bar{x})\frac{\sum_{j=1}^{n}(x_j - \bar{x})y_j}{\sum_{k=1}^{n}(x_k - \bar{x})^2}.$$

Substituting $\bar{y} = \frac{1}{n}\sum_{j=1}^{n} y_j$, we recast this expression as

$$\hat{y}_i = \sum_{j=1}^{n}\left\{\frac{1}{n} + \frac{(x_i - \bar{x})(x_j - \bar{x})}{\sum_{k=1}^{n}(x_k - \bar{x})^2}\right\} y_j. \tag{4}$$

The expression in braces involves only values of x; it does not depend on any y-values.

To simplify notation, we let h_{ij} denote the expression in braces and rewrite equation (4) as

$$\hat{y}_i = \sum_{j=1}^{n} h_{ij}y_j. \tag{5}$$

Thus h_{ij} indicates how changing y_j affects $\hat{y}_i$. In this and other regression models the (symmetric) $n \times n$ matrix $\mathbf{H} = (h_{ij})$ is known as the *hat matrix* because it takes the vector of observed y-values $(y_1, y_2, \ldots, y_n)^T$ into the vector of fitted y-values $(\hat{y}_1, \hat{y}_2, \ldots, \hat{y}_n)^T$ and thus puts hats on the y's. Ordinarily one pays particular attention to the observed and fitted y-values at the same data point. The notion of leverage formalizes this step. Specifically:

In linear regression the *leverage* of observation i is h_{ii}, the corresponding diagonal element of the hat matrix.

For the simple regression line

$$h_{ii} = \frac{1}{n} + \frac{(x_i - \bar{x})^2}{\sum_{k=1}^{n}(x_k - \bar{x})^2}. \qquad (6)$$

The hat matrix and its diagonal elements have several basic properties (Hoaglin and Welsch 1978):

- **H** is a projection matrix and hence idempotent ($\mathbf{H}^2 = \mathbf{H}$).

- $0 \leq h_{ii} \leq 1$ in general, and $h_{ii} \geq 1/n$ if the model includes a constant term.

- In simple linear regression $\sum_{i=1}^{n} h_{ii} = 2$ (unless all the x_i are equal), and hence the average size of an h_{ii} is $2/n$.

- More generally, in full-rank multiple regression with p explanatory variables (counting the constant if present), the average size of an h_{ii} is p/n.

- If $h_{ii} = 0$ or $h_{ii} = 1$, then $h_{ij} = 0$ for all $j \neq i$.

We interpret the two extreme cases $h_{ii} = 0$ and $h_{ii} = 1$ as follows. If $h_{ii} = 0$, then $\hat{y}_i$ must be 0; it is not affected by y_i or by the other y_j. A point with $x = 0$ when fitting a straight line through the origin provides a simple example. If $h_{ii} = 1$, then $\hat{y}_i = y_i$, and the regression model always fits this data point exactly. For example, sometimes one adds an indicator variable (which is 1 for observation i, and 0 elsewhere) to remove an outlier.

In simple regression the observations with highest leverage are precisely those whose x values are farthest from the mean of the observed x values.

Equation (6) shows that in simple regression the observations with highest leverage are precisely those with x_i farthest from $\bar{x}$. An observation has high leverage when it singlehandedly accounts for a large enough fraction of $\sum_{k=1}^{n}(x_k - \bar{x})^2$ or (rarely) when n is small enough. Thus we can discover whether any observations have

high leverage by plotting the data and looking for deviant x-values. For example, Figure 1 leads us to observation 1, whose leverage is 0.819 (the slope of the line in Figure 4).

What constitutes high leverage? A simple and convenient criterion comes from Huber (1981, pp. 160–162):

> Values $h_{ii} \leq 0.2$ appear to be safe, values between 0.2 and 0.5 are risky, and if we can control the design at all, we had better avoid values above 0.5.

To motivate these cutoffs in a way that applies to more general regression models, we begin with the fact that, if each y_j has variance σ^2 and (as is customary) the x_j are fixed, then the variance of $\hat{y}_i$ is $\sigma^2 h_{ii}$. We approach the variability of $\hat{y}_i$ by asking:

> How large would a sample of m independent observations, each with variance σ^2, have to be in order for its mean to have the same variance as $\hat{y}_i$?

The variance of the mean of that hypothetical sample is σ^2/m. Setting $\sigma^2/m = \sigma^2 h_{ii}$ yields $m = 1/h_{ii}$. That is, when we consider the variance of $\hat{y}_i$, we may think of h_{ii} as a reciprocal sample size. Thus, $h_{ii} > 0.5$ indicates that $\hat{y}_i$ is more variable than a mean of two independent observations, whereas $h_{ii} \leq 0.2$ ensures that $\hat{y}_i$ is no more variable than a mean of five observations. In simple regression the average of the h_{ii} is $2/n$. Such a value of h_{ii} corresponds to $m = n/2$, so that $\hat{y}_i$ would behave like a mean of $n/2$ independent observations. Ordinarily we prefer to avoid fitted values that behave as if they are based on only a small part of the data. On average, such observations become less likely as n increases, because the average of the h_{ii} decreases as $1/n$.

These guidelines aid in identifying, for further investigation, observations that have high leverage. Still, no single criterion seems to handle all patterns of leverage. Thus it is usually desirable to examine all the h_{ii} as a batch and to see whether any stand out.

For convenience, the remainder of this discussion uses the abbreviated notation h_i instead of h_{ii}.

Table 2 gives the numerical values of h_i in the data on Public Affairs Activities. No matter what cutoff we use to identify high-leverage points, observation 1 stands

i	h_i
1	.819
2	.085
3	.139
4	.093
5	.090
6	.099
7	.106
8	.092
9	.109
10	.122
11	.117
12	.128

Table 2: Values of leverage h_i in the Public Affairs Activities example.

out, and all the other observations have modest leverage (in fact, less than the average value, 2/12). It is reassuring to see in the h_i the same messages that are evident in Figure 1: x_1 is much larger than the other x-values, which spread over an interval.

In some instances the regression coefficients are of interest. For example, in the data on Public Affairs Activities we may interpret the slope as a typical per-person salary. To express b_2 as a linear function of the y_i, we recast equation (3) as

$$b_2 = \sum_{i=1}^{n} c_i y_i,$$

where $c_i = (x_i - \bar{x})/\sum_{k=1}^{n}(x_k - \bar{x})^2$ depends only on the x-values. (A similar relationship holds separately for each coefficient in a multiple regression model.) The c_i offer a straightforward quantitative way of showing the relative impact of the individual y-values on b_2 and of emphasizing that every y_i (with $x_i \neq \bar{x}$) has some effect on b_2.

Residuals

A thorough regression analysis should include examination of residuals, and simple linear regression is a good place to start building this beneficial habit. By definition, the *residual* for observation i is

$$e_i = y_i - \hat{y}_i = y_i - \bar{y} - b_2(x_i - \bar{x}). \tag{7}$$

We usually estimate σ^2, the variance of an individual

y-value about its mean, by the *residual mean square*

$$s^2 = \frac{1}{n-2} \sum_{i=1}^{n} e_i^2. \tag{8}$$

In checking for nonlinearity and nonconstant variability the customary approach plots e_i versus x_i and $\hat{y}_i$. These straightforward displays are often useful, but we do not discuss them further in this chapter.

To look at residuals effectively, we note that they usually do not all have the same variance, even though $\text{var}(y_i) = \sigma^2$ for each i. Indeed, in terms of the diagonal elements of the hat matrix,

$$\text{var}(e_i) = \sigma^2(1 - h_i). \tag{9}$$

Thus, not only does a large h_i mean that e_i will tend to be small; it also means that e_i has a correspondingly small variance. The standard deviation of e_i is $\sigma\sqrt{1 - h_i}$, which we estimate by $s\sqrt{1 - h_i}$. Dividing e_i by its estimated standard deviation (also known as its *estimated standard error*) yields the *standardized residual*

$$\frac{e_i}{s\sqrt{1 - h_i}}. \tag{10}$$

We work with the standardized residuals as if each of them has mean 0 and standard deviation 1; we can more easily identify large residuals (in magnitude) on this familiar scale.

A thorough regression analysis should include examination of residuals. To look at residuals effectively, we note that they usually do not all have the same variance.

For the example of Public Affairs Activities in U. S. Government agencies, Table 3 gives x_i, y_i, $\hat{y}_i$, e_i, h_i, $s\sqrt{1 - h_i}$, and the standardized residual. To check on the effect of standardizing the residuals, we look first at the estimated standard error $s\sqrt{1 - h_i}$. Each residual has a standard error of approximately 4, except for e_1, whose standard error is 1.8 because of the high leverage of (x_1, y_1). The standardized residuals in the last column call attention only to point 2, which has low leverage and a high y-value. We return to this data point after further diagnosis.

i	x_i	y_i	$\hat{y}_i$	e_i	h_i	$s\sqrt{1-h_i}$	$e_i/s\sqrt{1-h_i}$
1	1486	24.51	25.69	-1.18	.819	1.82	$-.65$
2	388	21.00	8.60	12.40	.085	4.10	3.02
3	650	11.47	12.68	-1.21	.139	3.98	$-.30$
4	202	5.80	5.71	.09	.093	4.08	.02
5	446	5.66	9.50	-3.84	.090	4.09	$-.94$
6	164	5.68	5.11	.57	.099	4.07	.14
7	128	5.24	4.55	.68	.106	4.05	.17
8	208	4.50	5.80	-1.30	.092	4.09	$-.32$
9	117	2.91	4.38	-1.47	.109	4.05	$-.36$
10	69	2.46	3.63	-1.18	.122	4.02	$-.29$
11	85	2.30	3.88	-1.58	.117	4.03	$-.39$
12	47	1.31	3.29	-1.98	.128	4.00	$-.49$

$\hat{y} = 7.74 + 0.0156(x - 332.5)$

$s = 4.29$

Table 3: Calculations leading to the standardized residuals for the Public Affairs Activities data. The values of y_i and related quantities are in millions of dollars.

Measuring an Observation's Influence by Deleting It

An observation's leverage indicates the potential impact that its y-value can have, entirely as a function of how its x-value relates to the other x-values in the data. We must probe further, however, to judge the actual impact of the observation. One straightforward approach is to delete it and see how various results change. We regard an observation as *influential* if deleting it produces a substantial change in an aspect of the regression that we consider important. What constitutes "a substantial change" is somewhat subjective. We usually evaluate an influence diagnostic at each observation and ask whether any values stand out. The deletion approach to diagnosis becomes quite powerful once we discover that simple formulas yield all the necessary quantities without actually removing each observation in turn and redoing the regression calculations with the reduced data sets.

A basic set of diagnostics reveals influence on $\hat{y}_i$, on the coefficients of the regression line, and on s^2. For each, we motivate the diagnostic measure and then present a short-cut formula (that uses h_i, e_i, and other quantities), skipping the details of the intervening algebra. (See Atkinson (1985), Belsley, Kuh, and Welsch (1980), Chatterjee and Hadi (1988), and Cook and Weisberg (1982) for derivations and discussions of these and other regression diagnostics.) One can reasonably introduce these measures in the context of simple linear regression and later extend them to multiple regression. In this way students encounter the ideas first in a familiar setting where the plot of y versus x conceals nothing.

To denote a quantity calculated from the data without observation i, we append "(i)" (read "not i" or "i omitted"). For example, $\hat{y}_i(i)$ is the predicted y-value at x_i when the regression is fitted to the data without observation i.

Influence on fitted values. To judge the influence of the observation (x_i, y_i) on the fitted value $\hat{y}_i$, we divide the difference between $\hat{y}_i$ and $\hat{y}_i(i)$ by an estimate of the standard error of $\hat{y}_i$. In the notation of Belsley, Kuh, and Welsch (1980), as revised by Velleman and Welsch (1981), the result is

$$\text{DFITS}_i = \frac{\hat{y}_i - \hat{y}_i(i)}{\text{s.e.}(\hat{y}_i)}. \quad (11)$$

(The acronym DFITS comes from Difference in FIT, Scaled.) To develop an estimate of the standard error of $\hat{y}_i$, we recall that $\text{var}(\hat{y}_i) = \sigma^2 h_i$. Because observation i may be anomalous, we use $s^2(i)$ instead of s^2 to estimate σ^2. Thus s.e.$(\hat{y}_i) = s(i)\sqrt{h_i}$. We easily calculate $s^2(i)$ in simple regression from the short-cut formula (Belsley, Kuh, and Welsch 1980, p. 64):

$$(n-3)s^2(i) = (n-2)s^2 - \frac{e_i^2}{1-h_i}. \quad (12)$$

A few algebraic substitutions in the numerator of equation (11) yield

$$\hat{y}_i - \hat{y}_i(i) = \frac{h_i e_i}{1 - h_i}, \tag{13}$$

and hence the short-cut formula

$$\text{DFITS}_i = \frac{\sqrt{h_i} e_i}{s(i)(1 - h_i)}. \tag{14}$$

We usually plot DFITS_i against i and look for unusual points, either positive or negative, which we then investigate as influential observations.

Equation (14) indicates that an observation can be influential because it has high leverage and a moderate residual or because it has moderate leverage and a large residual. Many combinations of h_i, e_i, and $s(i)$ yield the same value of DFITS_i. We emphasize that neither h_i alone nor e_i alone suffices to identify an influential observation.

An observation can be influential because it has high leverage and a moderate residual or because it has moderate leverage and a large residual.

Influence on regression coefficients. Some regression analyses focus on the values of the regression coefficients. Here also we can measure influence by deletion. To judge the influence of observation i on b_2, we form $b_2 - b_2(i)$ and divide this difference by an estimate of the standard deviation (standard error) of b_2:

$$\frac{b_2 - b_2(i)}{\text{s.e.}(b_2)}.$$

A moderate amount of algebra yields

$$b_2 - b_2(i) = \frac{x_i - \bar{x}}{\sum (x_k - \bar{x})^2} \times \frac{y_i - \hat{y}_i}{1 - h_i}$$

and we obtain $\text{s.e.}(b_2)$ from the standard formula

$$\text{var}(b_2) = \sigma^2 / \sum (x_k - \bar{x})^2.$$

To estimate σ^2 we again use $s^2(i)$ instead of s^2. Thus, in the notation of Belsley, Kuh, and Welsch (1980) we have

$$\begin{aligned}
\text{DBETAS}_{i2} &= \frac{b_2 - b_2(i)}{[s^2(i)/\sum (x_k - \bar{x})^2]^{1/2}} \\
&= \frac{(x_i - \bar{x}) e_i}{s(i)(1 - h_i) [\sum (x_k - \bar{x})^2]^{1/2}} \tag{15}
\end{aligned}$$

(The acronym DBETAS comes from Difference in BETA, Scaled.)

Similarly, to judge the influence of observation i on b_1, we divide $b_1 - b_1(i)$ by the estimate of s.e.(b_1) based on $s(i)$. The short-cut formula is

$$\text{DBETAS}_{i1} = \frac{e_i}{s(i)(1 - h_i)\sqrt{n}}. \tag{16}$$

We look at the values of $\text{DBETAS}_{ij}, i = 1, \ldots, n$, often by plotting DBETAS_{ij} against i for each j, and see whether any of them stick out from the rest.

Influence on s^2. Because s^2 plays an important role in tests and confidence intervals associated with the regression, it is useful to know whether any one observation influences s^2 (by inflating it). To diagnose this, we use $s^2(i)$, the residual mean square from the regression without observation i, calculated from equation (12). We plot $s^2(i)/s^2$ against i and give most attention to values noticeably smaller than 1.

Studentized residuals. We complete this set of diagnostic techniques by looking for observations that seem not to fit well with the model. The deletion approach leads us to consider the *predicted residual*, $y_i - \hat{y}_i(i)$. To rewrite this expression in terms of e_i and h_i, we use equation (13):

$$\begin{aligned}
y_i - \hat{y}_i(i) &= (y_i - \hat{y}_i) + (\hat{y}_i - \hat{y}_i(i)) \\
&= e_i + \frac{h_i e_i}{1 - h_i} = \frac{e_i}{1 - h_i}.
\end{aligned}$$

Because $e_i/(1 - h_i)$ has variance $\sigma^2/(1 - h_i)$, we use $s(i)/\sqrt{1 - h_i}$ as the estimated standard error to obtain the (externally) *studentized residual*

$$e_i^* = \frac{e_i}{s(i)\sqrt{1 - h_i}}. \tag{17}$$

By using $s(i)$ this definition focuses more attention on residuals that are large enough, relative to $\sqrt{1 - h_i}$, to inflate s. Thus we ordinarily use studentized residuals instead of standardized residuals (equation (10)).

We may examine the studentized residuals in a variety of ways, including a plot of e_i^* against i. The choice of $s(i)$ in the estimated standard deviation of $y_i - \hat{y}_i(i)$ yields e_i^* that individually follow a Student's t distribution on $n-3$ degrees of freedom. When we judge the extremeness of some e_i^* in terms of that distribution, however, we must make allowance for the fact that the largest $|e_i^*|$ will attract our attention first.

i	DFITS_i	DBETAS_{i1}	DBETAS_{i2}	$s^2(i)$	e_i^*
1	-1.338	$-.427$	-1.268	19.56	$-.63$
2	2.984	2.954	.422	1.76	9.79
3	$-.117$	$-.090$	$-.074$	20.24	$-.29$
4	.007	.007	$-.002$	20.42	.02
5	$-.294$	$-.282$	$-.083$	18.62	$-.93$
6	.044	.040	$-.018$	20.39	.13
7	.055	.049	$-.026$	20.37	.16
8	$-.096$	$-.092$	.029	20.22	$-.30$
9	$-.121$	$-.106$	.059	20.16	$-.35$
10	$-.104$	$-.086$	.058	20.25	$-.28$
11	$-.137$	$-.116$	.074	20.11	$-.38$
12	$-.182$	$-.147$	.108	19.93	$-.47$

Table 4: Selected regression diagnostics and related quantities for the Public Affairs Activities data.

When we compare equation (17) to equation (14), we see that both e_i^* and DFITS_i involve e_i, h_i, and $s(i)$. The two measures differ in how they handle the contribution of leverage. Specifically,

$$\text{DFITS}_i = e_i^* \left(\frac{h_i}{1 - h_i} \right)^{1/2}. \qquad (18)$$

Thus, for example, a moderate $|e_i^*|$ at a high-leverage observation produces a large $|\text{DFITS}_i|$, whereas a small enough h_i can offset a large $|e_i^*|$ and give a moderate $|\text{DFITS}_i|$.

In using influence diagnostics we emphasize plots and other displays. Although confidence in interpreting them may come only with experience, it is important to appreciate the element of judgment in data analysis. The plots and displays serve to focus attention on observations that should receive further scrutiny, in the hope of uncovering a concrete reason for their greater influence.

Example of Influence Diagnostics in Simple Linear Regression

We return to the example of employees involved in Public Affairs Activities. From Figure 1 and Table 2, the point (x_1, y_1) for the Department of Defense (DOD) has high leverage, so we pay particular attention to its influence. Table 4 presents the values of DFITS_i, DBETAS_{i1}, DBETAS_{i2}, $s^2(i)$, and e_i^*, and we next plot each of these against i.

The plot of DFITS_i in Figure 5 shows that point 1 (DOD) substantially lowers its fitted value, whereas

point 2 (HEW) has an even larger impact in the opposite direction. No other points seem noteworthy. The newspaper article does not indicate why the cost at HEW is so much greater than one would expect from the data on the other agencies. Focusing on the coefficients separately, the plot of DBETAS_{i1} in Figure 6 shows that point 2 contributes very substantially to raising the level of the line. DBETAS_{i2} (in Figure 7) confirms the impression that point 1 has a large impact in decreasing the slope of the line, as a consequence of its leverage. Also, point 2 plays a more modest, but still noticeable, role in increasing the slope of the line. Figure 8, showing $s^2(i)/s^2$, emphasizes that point 2 contributes a major share of the residual sum of squares (from Table 3, $e_2 = 12.40$). After further investigation we should consider setting aside the data on HEW. Finally, the plot of e_i^* in Figure 9 again shows how dramatically point 2 departs from the line that reasonably well summarizes the relationship between cost and number of employees in the other eleven agencies.

Comparing the DFITS_i and the e_i^* in Table 4, we see that, although e_1^* is modest, h_1 is so large that DFITS_1 leads us to flag the point as influential. For the other influential observation, e_2^* is so large that DFITS_2 is also large, despite the small value of h_2.

In summary, these influence diagnostics have led us to regard the data on DOD and HEW as influential and to consider the role that they may play in conclusions that we draw from these data.

Even though these measures of influence primarily rein-

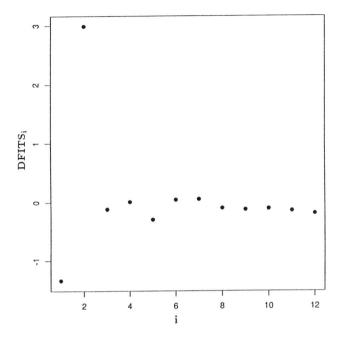

Figure 5: DFITS$_i$ versus i, showing influence on fitted value.

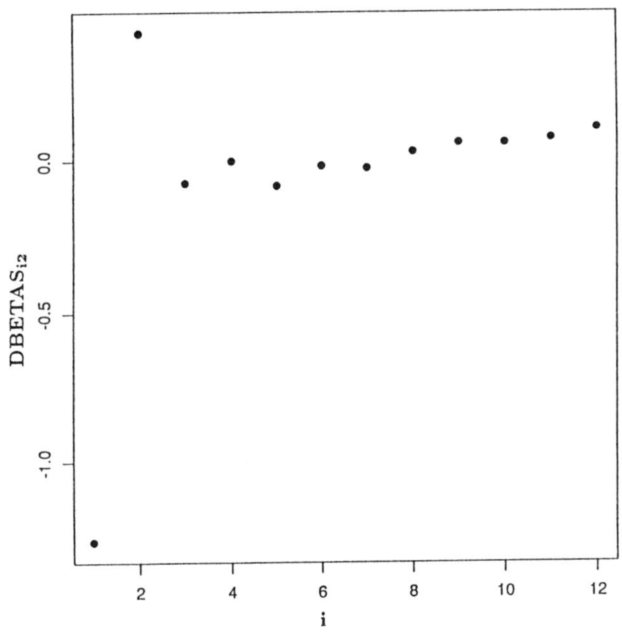

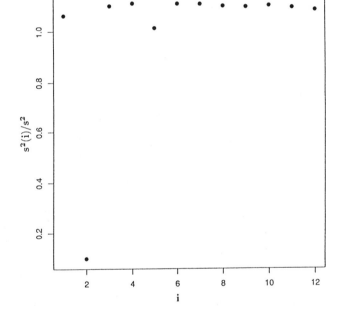

Figure 7: DBETAS$_{i2}$ versus i, showing influence on the slope.

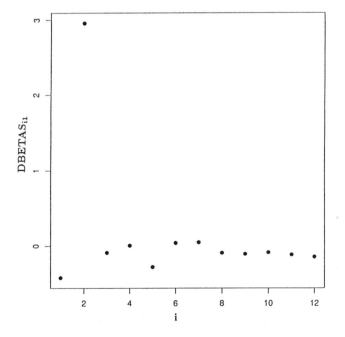

Figure 6: DBETAS$_{i1}$ versus i, showing influence on the constant term.

Figure 8: Plot of $s^2(i)/s^2$ versus i, showing that point 2 substantially inflates the residual mean square.

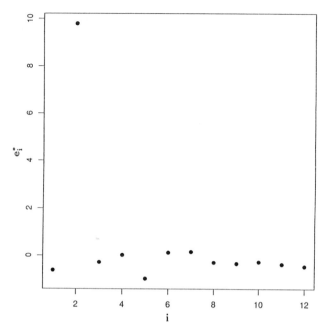

Figure 9: (Externally) studentized residual e_i^* versus i.

force messages that one can get from careful study of the plot of y versus x in Figure 1, it is worthwhile to look at them for two main reasons. First, they consolidate information about particular forms of influence. Second, they provide valuable preparation for diagnosing influence in multiple regression, where no single plot tells the whole story.

3. LEVERAGE AND INFLUENCE IN MULTIPLE REGRESSION

Diagnosis of influential observations in multiple regression typically relies more heavily on numerical measures of influence. Even a full set of scatterplots (y versus each x-variable and all the x-variables pairwise) may fail to reveal some types of influential data. Fortunately the diagnostic measures presented earlier extend readily to multiple regression. The books by Atkinson (1985), Belsley, Kuh, and Welsch (1980), Chatterjee and Hadi (1988), and Cook and Weisberg (1982), along with articles such as Hoaglin (1988), Hoaglin and Welsch (1978), and Velleman and Welsch (1981), give details and background of these and other regression diagnostics.

In matrix notation the multiple regression model

$$\mathbf{y} = \mathbf{X}\boldsymbol{\beta} + \boldsymbol{\varepsilon}$$

relates the *response variable* or *dependent variable*, whose observations form the $n \times 1$ vector $\mathbf{y}$, to the *carriers* or *explanatory variables* $X_1, \ldots, X_p$, which form the columns of the $n \times p$ matrix $\mathbf{X}$, via the $p \times 1$ vector of *regression coefficients* $\boldsymbol{\beta}$. Usually the X_j include the constant carrier, and we assume that $\mathbf{X}$ has full rank p. The *fluctuation* or *disturbance* vector $\boldsymbol{\varepsilon}$ has $\mathrm{E}(\boldsymbol{\varepsilon}) = \mathbf{0}$ and $\mathrm{cov}(\boldsymbol{\varepsilon}) = \sigma^2 \mathbf{I}$ and is often assumed to have an n-variate normal distribution.

To denote the ith row of $\mathbf{X}$, we use the row vector $\mathbf{x}_i$.

The least-squares estimate of $\boldsymbol{\beta}$ is

$$\mathbf{b} = (\mathbf{X}^T\mathbf{X})^{-1}\mathbf{X}^T\mathbf{y}, \qquad (19)$$

and the residual vector is

$$\mathbf{e} = \mathbf{y} - \mathbf{X}\mathbf{b}.$$

The residual mean square $s^2 = \mathbf{e}^T\mathbf{e}/(n-p)$ is an unbiased estimate of σ^2.

Substituting $\mathbf{b}$ from equation (19) in the vector of fitted values

$$\hat{\mathbf{y}} = \mathbf{X}\mathbf{b},$$

we get

$$\hat{\mathbf{y}} = \mathbf{X}(\mathbf{X}^T\mathbf{X})^{-1}\mathbf{X}^T\mathbf{y},$$

and thereby expose the linear relationship between $\hat{y}_i$ and the y_j. To abbreviate this as $\hat{\mathbf{y}} = \mathbf{H}\mathbf{y}$, we define the *hat matrix*,

$$\mathbf{H} = \mathbf{X}(\mathbf{X}^T\mathbf{X})^{-1}\mathbf{X}^T. \qquad (20)$$

We note that $\mathbf{H}$ projects $\mathbf{y}$ onto the space spanned by the columns of $\mathbf{X}$. The elements of $\mathbf{H}$ have the same interpretation as in simple regression. The average of the h_i is p/n, and Huber's cutoffs at 0.2 and 0.5 still apply.

For the influence of observation i on $\hat{y}_i$, we have

$$\mathrm{DFITS}_i = \frac{\hat{y}_i - \hat{y}_i(i)}{\mathrm{s.e.}(\hat{y}_i)}$$
$$= \frac{\hat{y}_i - \mathbf{x}_i\mathbf{b}(i)}{s(i)\sqrt{h_i}},$$

as well as the short-cut formula

$$\mathrm{DFITS}_i = \frac{\sqrt{h_i}e_i}{s(i)(1-h_i)},$$

where $s^2(i)$ comes from

$$(n-p-1)s^2(i) = (n-p)s^2 - \frac{e_i^2}{1-h_i}.$$

When we look at the influence of observation i on b_j, the estimate of β_j, we use the general definition

$$\text{DBETAS}_{ij} = \frac{b_j - b_j(i)}{\text{s.e.}(b_j)}$$
$$= \frac{b_j - b_j(i)}{s(i)\sqrt{(\mathbf{X}^T\mathbf{X})_{jj}^{-1}}}.$$

and calculate $b_j - b_j(i)$ from the vector formula

$$\mathbf{b} - \mathbf{b}(i) = \frac{(\mathbf{X}^T\mathbf{X})^{-1}\mathbf{x}_i^T e_i}{1 - h_i}.$$

For the (externally) studentized residual, the formula remains

$$e_i^* = \frac{e_i}{s(i)\sqrt{1 - h_i}};$$

and when the ε_i are jointly normal, each e_i^* has a t distribution on $n - p - 1$ degrees of freedom.

In multiple regression it is possible to display information on the relation of y to a single carrier X_j in a way that parallels the plot of y versus x in simple regression. We focus on each carrier X_j in turn, remove from X_j and y their regression on the rest of the carriers, and plot the resulting y-residuals against the corresponding X_j-residuals. To present the components more compactly, we let y_{*j} denote the residual from the multiple regression of y on all the carriers except X_j (i.e., $X_1, \ldots, X_{j-1}, X_{j+1}, \ldots, X_p$), and we let X_{j*j} denote the residual from the multiple regression of X_j on all the other carriers.

The *partial regression plot* (also known as an adjusted-variable plot or an added-variable plot) for X_j puts y_{*j} on the vertical axis and X_{j*j} on the horizontal axis. Because

$$y_{*j} = b_j X_{j*j} + e$$

(see, for example, Mosteller and Tukey 1977, p. 344), the plot enables us to see several key features at once:

1. The regression line through the origin has slope b_j, the least-squares estimate of β_j in the multiple regression.

2. Observations that exert substantial influence on b_j will stand out, by virtue of the leverage associated with their position on the X_{j*j} axis.

3. The residuals from the regression of y_{*j} on X_{j*j} are the same as those for the multiple regression, so they

are visible in the plot as the vertical deviations of the points from the line (which we sometimes include).

It is easiest to come to grips with regression diagnostics and other aspects of multiple regression in situations that have only two nonconstant carriers, because we can still see a lot in a relatively small number of plots. For example, a single scatterplot still shows the relation of the two nonconstant carriers, making it easy to check that the numerical values of the h_i reflect identifiable features of that plot. Experience with such examples helps in interpreting the diagnostic measures in regressions with more carriers.

Diagnosis of influential observations in multiple regression typically relies more heavily on numerical measures of influence.

To carry over what we are able to see in simple regression about how individual points affect the slope of the line, we reduce the two-variable problem to two one-variable problems by using the partial regression plot. The idea is that, to look at the regression of y on X_3 in the presence of X_1 (the constant) and X_2, we first clear the X_1 and X_2 components (that is, the simple regression on X_2) out of both y and X_3. We interchange the roles of X_2 and X_3 to look at the regression of y on X_2 in the presence of X_1 and X_3. Then we continue with the influence diagnostics DBETAS_{ij}, DFITS_i, and $s^2(i)$ and the studentized residuals e_i^*.

Routine use of even the limited set of diagnostics presented here should give close contact with the features of a set of multiple regression data. Diagnostics for the influence of individual observations provide a solid starting point for understanding approaches that seek to reveal groups of observations that are jointly influential but not individually influential — a challenging area in which research continues.

4. DEPARTURES FROM ASSUMPTIONS

A variety of standard diagnostics aim to reveal ways in which a description or model fails to capture the behavior of the data. As a leading example, the classical plot of the residual, e_i, against the fitted value, $\hat{y}_i$, in a regression can show nonconstant variability and nonlinearity, as well as other features. More recent developments have refined this plotting strategy by focusing on ways of summarizing such patterns in the residuals and making tentative inferences about changes in the model (e.g., the most promising transformations for stabilizing the variability lie in a certain range of powers).

Constructed Variables

Many of these techniques involve constructing one or more special variables and then regressing the residuals on these constructed variables. We illustrate this process by diagnosing the presence of systematic nonadditivity in a two-way layout with one observation per cell. This structured type of data has I rows, J columns, and one observed value of the response, y_{ij}, for each combination of the row index i and the column index j, which we call the (i, j) cell.

The customary analysis of a two-way layout uses the simple additive description

$$y_{ij} = m + a_i + b_j + e_{ij} \qquad (21)$$

to decompose the y_{ij} in terms of a common value (m), row effects $(a_i, i = 1, \ldots, I)$, column effects $(b_j, j = 1, \ldots, J)$, and residuals $(e_{ij}, i = 1, \ldots, I, j = 1, \ldots, J)$. One standard set of constraints, appropriate when we determine m, the a_i, and the b_j by least squares, requires that $\sum a_i = 0, \sum b_j = 0, \sum_j e_{ij} = 0$ for each i, and $\sum_i e_{ij} = 0$ for each j. We could write this least-squares analysis as a multiple regression in which the entries of $\mathbf{X}$ are either 0, $+1$, or -1; but we do not need the details here, and we omit them. Instead, we use the notation • in place of a subscript i or j to indicate the result of averaging over that subscript, and thus we can write m, a_i, and b_j in terms of the y_{ij}:

$$
\begin{aligned}
m &= y_{\bullet\bullet} \\
a_i &= y_{i\bullet} - y_{\bullet\bullet} \qquad (22) \\
b_j &= y_{\bullet j} - y_{\bullet\bullet}.
\end{aligned}
$$

Any departure of the data from additivity shows up in the residuals,

$$
\begin{aligned}
e_{ij} &= y_{ij} - \hat{y}_{ij} \\
&= y_{ij} - (m + a_i + b_j) \\
&= y_{ij} - (y_{i\bullet} + y_{\bullet j} - y_{\bullet\bullet}).
\end{aligned}
$$

Ideally, the additive description is satisfactory, and the e_{ij} reflect only chance fluctuations. Sometimes, however,

the e_{ij} show a systematic pattern. To diagnose the most common patterns of nonadditivity, Tukey (1977, Section 10F) constructed the *comparison values*

$$cv_{ij} = \frac{a_i b_j}{m}$$

(for data in which $m \neq 0$) and introduced the diagnostic plot of e_{ij} versus cv_{ij}. That is, the cv_{ij} are the values of the constructed variable.

When this diagnostic plot resembles a straight line, one can consider adding a multiple of the comparison values to the fit. Such a pattern of nonadditivity, however, often suggests that we would do better to analyze the data in a transformed scale, in which the data would show much smaller (perhaps negligible) departures from additivity. A slope of r in the diagnostic plot suggests that we transform each y_{ij} by the $1 - r$ power or some nearby simple power (if this makes sense in the context of the data):

$$
\begin{aligned}
y_{ij} &\rightarrow y_{ij}^{1-r} \quad \text{if } r \neq 1 \\
y_{ij} &\rightarrow \log y_{ij} \quad \text{if } r = 1.
\end{aligned}
$$

Emerson (1983) discusses the background of the plot and the basis for this interpretation.

As an example of this approach to diagnosing systematic nonadditivity, Table 5 shows a two-way layout and its decomposition. These data are the rates (per appearance) that the monthly magazine of a professional society once charged for display advertising. The two factors are the size of the advertisement and the number of times that the ad was to appear. In panel (b) the strong systematic pattern of the residuals (positive in the upper left and lower right portions and negative in the lower left and upper right portions, with magnitudes increasing away from the center) announces that the additive summary is far from satisfactory. The diagnostic plot of residuals versus comparison values (Figure 10) shows very nearly a straight line whose slope differs little from 1.

As this slope suggests, after we transform the data to the logarithmic scale, the simple additive model fits very well (but not perfectly). We may transform this description back into a multiplicative description in the original scale: $\hat{y}_{ij} = M \times A_i \times B_j$. In this example it may be helpful to take $A_1 = 1$ and $B_1 = 1$. Then the other A_i indicate how the costs for fractional-page ads relate to the cost of a full-page ad, and the other B_j summarize the discounts available for running an ad multiple times.

When the data may contain anomalous observations, it

a. The Data (in dollars per appearance)

Size	Number of times to appear			
	1	4	6	12
Full page	3000	2850	2650	2500
2/3 page	2400	2280	2120	2010
1/2 page	1896	1800	1674	1590
1/3 page	1460	1387	1290	1225
1/4 page	1095	1040	970	920
1/6 page	799	759	705	670

b. The Decomposition

Size	Number of times to appear				a_i
	1	4	6	12	
Full page	103.75	42.75	−39.42	−107.08	1121.25
2/3 page	51.25	20.25	−21.92	−49.58	573.75
1/2 page	9.75	2.75	−5.42	−7.08	111.25
1/3 page	−26.75	−10.75	10.08	27.42	−288.25
1/4 page	−57.50	−23.50	24.33	56.67	−622.50
1/6 page	−80.50	−31.50	32.33	79.67	−895.50
b_j	146.25	57.25	−60.58	−142.92	1628.75

Table 5: A 6×4 two-way layout of rates that a monthly professional magazine charged for display advertising, along with its additive decomposition into common value (m), row effects (a_i), column effects (b_j), and residuals (e_{ij}). In the decomposition the row effects, column effects, and common value border the table of residuals.

is often more effective to base a diagnostic plot for non-additivity on a resistant fit (Section 5). The reason lies in the phenomenon of *leakage*: the observed value at one data point affects ("leaks into") the fitted values at all the other data points. In a two-way layout, trouble from a sour value in one cell leaks into the fitted values in the other cells, as we now demonstrate. By writing the least-squares fit from equations (21) and (22) as

$$\hat{y}_{ij} = y_{i\bullet} + y_{\bullet j} - y_{\bullet\bullet}$$

and denoting an element of the hat matrix as $h_{ij,kl}$, so that

$$\hat{y}_{ij} = \sum_{k=1}^{I} \sum_{l=1}^{J} h_{ij,kl} y_{kl},$$

we may calculate the $h_{ij,kl}$ directly:

$$
\begin{aligned}
h_{ij,ij} &= (I + J - 1)/IJ; \\
h_{ij,il} &= (I - 1)/IJ, & l \neq j; \\
h_{ij,kj} &= (J - 1)/IJ, & k \neq i; \\
h_{ij,kl} &= -1/IJ, & k \neq i, l \neq j.
\end{aligned}
$$

Thus, because all the elements of the hat matrix are nonzero, an anomalous observation in one cell leaks into

the fitted values in all cells. As an extreme case, if $\{y_{ij}\}$ contains only one nonzero entry, the residuals coincide with the comparison values. Leakage allows the isolated observation to produce a pattern of residuals that looks exactly like systematic nonadditivity.

Probability Plotting

A well-developed body of graphical techniques aids in diagnosing departures from a specified form of distribution, such as the normal distribution. Briefly, one puts the data in numerical order and plots the ordered observations (or some smaller set of observed quantiles) against values that show the typical behavior of those quantiles for the given distribution. Usually the given distribution is actually a family of distributions in which the members all have the same shape but differ in location and scale. If all is well, the plot resembles a straight line whose slope reflects the scale of the data and whose intercept reflects the location. Departures from a roughly straight-line pattern reveal skewness, heavier tails, lighter tails, or individual outliers.

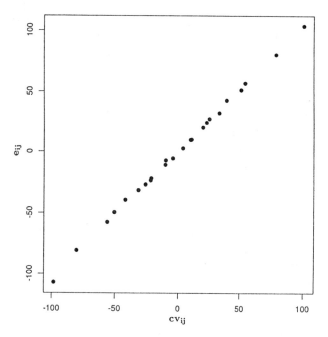

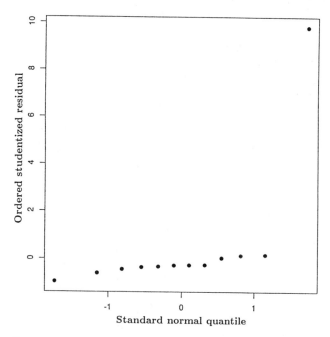

Figure 10: Plot of residuals e_{ij} versus comparison values $cv_{ij} = a_i b_j / m$ for the data on rates for display advertising. The slope close to 1 suggests transforming the data to the logarithmic scale.

Figure 11: Normal probability plot of the studentized residuals from the least-squares line for the data on Public Affairs Activities.

To develop the plotting values for the horizontal coordinate, we begin with a sample, $\{y_1, y_2, \ldots, y_n\}$, and denote the ordered sample (the sample *order statistics*) by $y_{(1)} \le y_{(2)} \le \cdots \le y_{(n)}$. The corresponding random variables are $Y_{(1)} \le Y_{(2)} \le \cdots \le Y_{(n)}$. When these are an ordered sample of n from the distribution whose cumulative distribution function is F, we plot $y_{(i)}$ against the median of the sampling distribution of $Y_{(i)}$, which we abbreviate as $\text{med}(Y_{(i)})$ (with i fixed). Conveniently, for any continuous distribution, $\text{med}(Y_{(i)})$ falls very close to the point where the value of F equals $(i - \frac{1}{3})/(n + \frac{1}{3})$ (see, for example, Hoaglin (1983)). To find that point, we use the inverse of the c.d.f.:

$$\text{med}(Y_{(i)}) \approx v_i = F^{-1}\left(\left(i - \frac{1}{3}\right) \middle/ \left(n + \frac{1}{3}\right)\right). \quad (23)$$

Thus the plot contains the points $(v_i, y_{(i)}), i = 1, \ldots, n$. Both coordinates are quantiles. Thus this plot belongs to a class of displays known as *quantile-quantile plots* (Q–Q plots). Wilk and Gnanadesikan (1968) discuss Q–Q plots and related techniques in greater generality.

When F is a member of a location-scale family of distributions (such as the normal distributions), it suffices to take the v_i in equation (23) from the standard member of the family (e.g., the normal distribution with mean 0

and standard deviation 1). The basis for this simplification lies in the relation between any two members, F_1 and F_2, of the family: For some constant μ and some positive constant σ, $F_2(y) = F_1((y - \mu)/\sigma)$ for all y. In terms of quantiles,

$$F_2^{-1}(p) = \mu + \sigma F_1^{-1}(p)$$

for $0 \le p \le 1$. Thus one common form of normal probability plot puts the quantiles of the standard normal distribution on the horizontal axis.

Although small samples offer very limited information on distribution shape, a normal probability plot of the residuals in a regression can reveal serious departures from the assumed normal distribution of the fluctuations. The residuals, however, are not an unstructured sample. From the role of the hat matrix in $\hat{\mathbf{y}} = \mathbf{H}\mathbf{y}$, we may write

$$\mathbf{e} = (\mathbf{I} - \mathbf{H})\mathbf{y} = (\mathbf{I} - \mathbf{H})\boldsymbol{\varepsilon}, \quad (24)$$

because $(\mathbf{I} - \mathbf{H})\mathbf{X}\boldsymbol{\beta} = \mathbf{0}$. For an individual residual we have

$$e_i = \varepsilon_i - \sum_{j=1}^{n} h_{ij}\varepsilon_j,$$

and in many regressions the second term on the right is more important than the first. When the distribution of the ε_i is not normal, such averaging tends to make the e_i more nearly normal than the ε_i, a phenomenon known as the "supernormality" of residuals.

Two other features of equation (24) also deserve consideration. First, $\text{var}(e_i) = \sigma^2(1 - h_i)$, so the residuals are ordinarily not equally variable. Second, because the covariance matrix of $\mathbf{e}$ is $\text{cov}(\mathbf{e}) = \sigma^2(\mathbf{I} - \mathbf{H})$, the residuals are correlated. Usually these correlations are not large enough to cause difficulty, and it is customary to ignore them. Many probability plots of residuals, however, do adjust for the unequal variances. For example, they may use the studentized residuals, e_i^* (equation (17)). Atkinson (1981; 1985, Section 4.2) presents a more refined plot based on the $|e_i^*|$.

To illustrate the simple probability plot of studentized residuals, we return to the data on Public Affairs Activities. Table 4 gives the e_i^*, and Figure 11 shows the probability plot. In this instance the plot is dominated by the point for the Department of Health, Education, and Welfare. Otherwise we see no indication of difficulty.

5. ROBUST/RESISTANT ALTERNATIVES

When data contain isolated anomalous observations, many aspects of their analysis would be simpler if no diagnosis were required. In part, this is the goal of *resistant* techniques: to summarize the behavior of the good observations and leave sizable residuals that lead the analyst directly to the anomalous observations. For example, among ways of summarizing the location of a single sample, the median (a value such that half the sample lies below it and the other half above) offers much resistance. One could replace nearly half the observations with arbitrarily large values without causing the median to become arbitrarily large. By contrast, the mean has no resistance at all. Even one highly deviant observation can pull the mean away from the good data. Although in this simple situation one could use either set of residuals to find the deviant observation, the residuals from the median would make it stand out. More generally, resistant techniques provide a cleaner separation of the data into fit and residuals.

The related concept of *robustness* focuses on departures from assumptions. If we are analyzing samples of data that we know come from normal distributions (a rare occurrence), then the sample mean, the sample variance, and Student's t provide the best estimates of the population mean and variance and the best tests or confidence intervals. In practice, however, data often come from some distribution that, although still symmetric, has heavier tails than the normal. Such heavier tails tend to make the sample mean more variable, and it may no longer be clear what the sample variance is estimating. Fortunately, as the discussion in Section 5 of Chapter 1 shows, Student's t suffers little adverse effect. It is reasonably robust to departures from normal data.

For estimating the population mean, as Chapter 9 explains, statisticians have devised a variety of robust techniques. The best of these are nearly as accurate as the sample mean when the data actually come from a normal distribution, and they perform nearly as well as the best procedure when the data come from any of several distributions with heavier tails. Chapters 9–11 of Hoaglin, Mosteller, and Tukey (1983) give one introduction to such robust techniques. Iglewicz (1983) discusses the special features that arise in estimating scale robustly, and Li (1985) provides an accessible treatment of robust regression.

> **Because they must cope with individual samples that contain deviant observations, robust techniques are generally resistant. Not all resistant techniques, however, are robust.**

Because they must cope with individual samples that contain deviant observations, robust techniques are generally resistant. Not all resistant techniques, however, are robust. The median is not adequately robust, because its value depends directly on only one or two observations at the center of the sample. Thus it is substantially more variable than the sample mean when the data are normal; and its performance may be inferior to the best procedure in other, heavier-tailed distributions.

Resistance in Regression

Like its least-squares cousin the mean, ordinary regression offers no resistance. We see this in the hat matrix, which usually has few zero entries anywhere. Thus, for example, as Figure 2 illustrates, simple regression allows a single observation with high leverage ($h_i = .8$) to seize control of the fitted line. In the actual data for that example, however, Tables 3 and 4 show that observation 2, which has the smallest leverage ($h_i = .085$) but a deviant y-value, produces the largest values of DFITS$_i$ and DBETAS$_{i1}$ and the second largest value (in magnitude) of DBETAS$_{i2}$. Thus, as long as an observation has any leverage at all (and $h_i \geq 1/n$ whenever the regression model includes an intercept term), a deviant y-value can distort the regression. The observed y_j affects the other fitted values, $\hat{y}_i$, through the off-diagonal elements of the hat matrix, h_{ij} ($i \neq j$) (see equation (5)). As Section 4 explains, we call this phenomenon leakage.

Ideally, a resistant regression should control leverage and leakage. As long as an observation follows the pattern of the rest of the data, it should contribute to the fit. But when the observation is anomalous, a resistant technique should prevent distortion and give a large residual. Thus the fitted values $\hat{y}$ can no longer be a linear function of the observed values y, as in linear regression ($\hat{y} = Hy$). For such situations we may generalize the definition of leverage to $\partial \hat{y}_i / \partial y_i$ with the understanding that the result will depend on y as well as on X. When observation i is anomalous, a resistant technique should arrange to have $\partial \hat{y}_i / \partial y_i = 0$ (i.e., beyond a certain point all change in observation i goes into the corresponding residual).

Least Median of Squares

Research in statistics has produced a variety of techniques for resistant regression. To illustrate, we now consider one of these, *least median of squares* (LMS), which uses a novel approach to give the greatest possible resistance.

Whereas least-squares fitting minimizes the *sum* of the squared residuals, $\sum_i e_i^2$, Rousseeuw (1984) introduced the idea of minimizing the *median* of the squared residuals, $\text{med}_i\{e_i^2\}$. That is, one finds the regression coefficients β that minimize

$$\text{med}_i\{(y_i - x_i\beta)^2\}.$$

The use of the median here allows deviant observations to produce large residuals instead of distorting the fit. Specifically, as long as $p > 1$ and any p of the observations uniquely determine β, this procedure can tolerate situations in which arbitrarily deviant observations replace nearly half the data without allowing β to become arbitrary. Rousseeuw and Leroy (1987) examine the properties of LMS in detail.

Applying the LMS approach to multiple regression problems involves substantially more computation than ordinary least squares. Timing studies reported by Rousseeuw (1984), however, indicate that, for $p \leq 10$ and $n \leq 200$, the computational effort does not become excessive. Some systems of statistical software now include LMS regression.

For the data on Public Affairs Activities (Table 1), the LMS line (with y in millions of dollars) is

$$0.96 + 0.0159x = 6.24 + 0.0159(x - 332.5).$$

The slope differs hardly at all from that of the least-squares line (Table 3); but the central value, 6.24, is 1.5 units lower, primarily because LMS does not allow observation 2 (HEW) to shift the line upward. For this deviant observation, the LMS line provides the desired resistance.

When we vary y_1 and study the slope of the LMS line, however, we find considerable sensitivity. Figure 12 (analogous to Figure 3) plots the slope against y_1. When y_1 is too small or too large, LMS ignores observation 1 and yields the line $-0.09 + 0.0340x$. In between, for $11.508 \leq y_1 \leq 80.508$, observation 1 essentially controls the slope of the line, which ranges from 0.0064 to 0.0558. Not surprisingly, the fitted value $\hat{y}_1$ exhibits similar behavior. Thus, although making y_1 arbitrarily extreme does not cause the slope or intercept of the LMS line to become arbitrary, the high leverage of observation 1 permits it to have substantial impact. (If the other 11 observations lay close to a single straight line, however, varying y_1 would have much less impact.) In keeping with the theme of this chapter, the key message is that, even when we use a resistant method, we need to appreciate the leverage associated with each observation.

6. COOPERATIVE DIVERSITY

As the diagnostic plot for nonadditivity and the discussion of probability plotting (both in Section 4) suggest, opportunities for diagnosis arise in settings other than regression. We now briefly discuss the "principle of cooperative diversity," which underlies a number of diagnostic techniques in non-regression settings.

In some structures of data we can learn more for diagnostic purposes by using different parts of the data (parts that are systematically related) than by leaving out individual observations in turn. Mainly, we have in mind data that may be well described by a few parameters, and we ask whether the diverse parts of the data seem to indicate different values of those parameters. Close agreement suggests that the chosen form of description is satisfactory. Lack of agreement may point toward a remedy. The diagnostic information comes from the cooperation among the parameter values calculated from the diverse parts of the data.

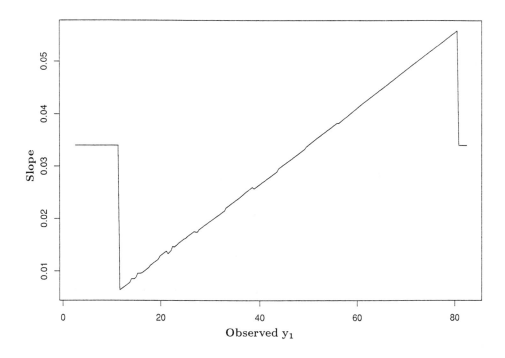

Figure 12: Slope of the LMS line versus y_1 for the Public Affairs Activities data.

To illustrate cooperative diversity, we use a single sample or distribution. The customary summaries focus on location (as described by the mean or the median) and spread (as described by the standard deviation or the interquartile range). Such a description is most helpful when the data are distributed more or less symmetrically about some central value. Thus we usually want to know about possible asymmetry or skewness. In practice we gain an initial appreciation of skewness from a stem-and-leaf display or a histogram of the data. For a quantitative indication of the presence and pattern of skewness, we look at a sequence of summary values from exploratory data analysis known as the *midsummaries* (Tukey 1977, Section 3E; Emerson and Stoto 1983).

The midsummaries are averages of pairs of *letter values*, a set of selected quantiles from either a sample or a theoretical distribution. The letter values begin with the median, which cuts the data or distribution into a lower half and an upper half (tail area $\frac{1}{2}$), and the lower and upper quartiles or fourths (tail area $\frac{1}{4}$), which further cut the lower half and upper half into quarters. The letter values continue outward into the tails by successively halving the tail area to produce the eighths, sixteenths, and so on. Thus the selected quantiles give considerable information on the tails. For the present purpose

we need not be concerned with the details of calculating the letter values of a set of data. We simply refer to them by a scheme of one-letter tags: M for the median, F for the fourths, E for the eighths, and so on in reverse order through the alphabet (D, C, B, A, Z, Y, X, ...). Except for the median, the letter values come in pairs (e.g., the lower and upper fourths, F_L and F_U).

For each pair of letter values, the corresponding midsummary is the average of the two letter values. We abbreviate it as "mid" followed by the letter used to tag the letter value. Thus, for example, the mid-fourth is

$$\text{midF} = \frac{1}{2}\left(F_L + F_U\right).$$

The sequence of midsummaries helps in diagnosing skewness. Here the different parts of the data are the letter values, systematically related through the choice of tail areas. In a perfectly symmetric set of data, all the midsummaries would be equal to the median. If the data were skewed to the right, the midsummaries would increase as they came from letter values further into the tails. For data skewed to the left, they would decrease. A sizable example shows how such a pattern arises in practice.

Tag	Lower letter value	Mid-summary	Upper letter value	Tail area
M		2.543		1/2
F	0.733	5.009	9.284	1/4
E	0.320	10.358	20.396	1/8
D	0.186	19.410	38.634	1/16
C	0.117	34.205	68.294	1/32
B	0.078	54.326	108.575	1/64
A	0.057	88.968	177.878	1/128
Z	0.042	140.049	280.057	1/256
Y	0.024	187.732	375.440	1/512
X	0.012	271.900	543.788	1/1024
W	0.004	294.864	589.723	1/2048
1	0.004	347.725	695.445	

Table 6: Letter values and midsummaries for the volume (in thousands of acre-feet) of 2509 oil and gas deposits. For completeness we include the extremes, the smallest and largest observations, with "1" as the tag.

In research on oil and gas deposits, geologists measure a number of attributes of the individual deposits in an area. Andreatta et al. (1988) examined the statistical distribution of several such measurements in data from the Lloydminster area of Alberta and Saskatchewan. For an expanded set of 2509 deposits, Table 6 gives letter values and midsummaries of the volume (in thousands of acre–feet). Because the midsummaries increase rapidly as the letter values move from the median into the tails (i.e., as the tail area decreases), this diagnostic information indicates that these data are strongly skewed to the right. Such a pattern often responds to a suitable transformation (e.g., replacing each observation by its logarithm) that renders the sample more nearly symmetric. One can readily try transformations that seem promising and then examine the midsummaries in each new scale. Emerson and Stoto (1983) devised a plot that uses the midsummaries to suggest the most promising transformations. In the present example the volumes are very nearly symmetric in the logarithmic scale.

This uncomplicated example illustrates the basic idea of working systematically with diverse parts of the data. In the original scale of volume the midsummaries told us that the data do not have a center of symmetry that could serve as a typical value. In the log scale the midsummaries would show close agreement.

Although not yet as widely applied as the ideas that underlie many regression diagnostics, the "principle of co-operative diversity" offers a potentially fruitful approach to a variety of situations. Hoaglin and Tukey (1985) list several further examples of cooperative diversity and apply the principle to develop two plots for diagnosis within a family of discrete distributions.

7. CONCLUSION

One brief chapter cannot provide comprehensive coverage of the varied approaches that underlie the rich array of diagnostic techniques. Thus the present account has focused primarily on the leading case of regression. The diagonal elements of the hat matrix, the DBETAS, the studentized residuals, and the partial regression plot illustrate an important attitude: Good data analysis involves at least as many numbers in output as one has in the original data. More generally, "leave out one" and other forms of sensitivity analysis provide a valuable way of constructing diagnostic measures in many situations.

In regression and other settings, constructed variables and probability plotting both have a variety of uses, extending well beyond the brief examples given here.

When available, resistant methods offer reduced sensitivity to anomalous observations, often making such data easier to detect. For systematic understanding of a data set, however, we still gain from applying both diagnostic techniques and resistant fitting.

Apart from regression, many structures of data do not lend themselves to "leave out one." For some of these it may be possible to designate suitable parts of the data and apply the principle of cooperative diversity. Several techniques for diagnosing aspects of distribution shape in discrete and continuous data arise in this way.

Good data analysis involves at least as many numbers in output as one has in the original data.

A relatively recent addition to the data analyst's toolkit, diagnostic techniques have done much to sharpen several types of analysis, especially multiple regression. We can expect a variety of further developments as research continues.

Acknowledgments

The author is grateful to John D. Emerson, Mark E. Glickman, David S. Moore, and Frederick Mosteller for helpful comments and suggestions.

Mark E. Glickman provided valuable assistance with the example in Section 5 and especially with the figures.

Preparation of this chapter was facilitated by grant SES–8908641 from the National Science Foundation to Harvard University.

David C. Hoaglin is a Research Associate in the Department of Statistics at Harvard University and also a Senior Scientist at Abt Associates Inc., an applied social research firm in Cambridge, MA. He received a B.S. in mathematics from Duke University in 1966 and a Ph.D. in statistics from Princeton University in 1971. Prior to his present positions he was on the faculty of the Statistics Department at Harvard from 1970 to 1977. His research interests include data analysis, robustness, statistical computing, statistical graphics, and applications of statistics to policy problems. He is a fellow of the American Statistical Association and the American Association for the Advancement of Science and an elected member of the International Statistical Institute.

REFERENCES

Andreatta, G., Kaufman, G. M., McCrossan, R. G., and Procter, R. M. (1988), "The Shape of Lloydminster Oil and Gas Deposit Attribute Data," in *Quantitative Analysis of Mineral and Energy Resources*, eds. C. F. Chung, A. G. Fabbri, and R. Sinding-Larsen, Dordrecht and Boston: D. Reidel, pp. 411–431.

Atkinson, A. C. (1981), "Two Graphical Displays for Outlying and Influential Observations in Regression," *Biometrika*, **68**, 13–20.

—— (1985), *Plots, Transformations and Regression: An Introduction to Graphical Methods of Diagnostic Regression Analysis*, Oxford and New York: Oxford University Press.

Belsley, D. A., Kuh, E., and Welsch, R. E. (1980), *Regression Diagnostics: Identifying Influential Data and Sources of Collinearity*, New York: John Wiley.

Chatterjee, S. and Hadi, A. S. (1986), "Influential Observations, High Leverage Points, and Outliers in Linear Regression," *Statistical Science*, **1**, 379–416 (with discussion).

—— (1988), *Sensitivity Analysis in Linear Regression*, New York: John Wiley.

Cook, R. D. and Weisberg, S. (1982), *Residuals and Influence in Regression*, New York and London: Chapman and Hall.

Emerson, J. D. (1983). "Mathematical Aspects of Transformation," in *Understanding Robust and Exploratory Data Analysis*, eds. D. C. Hoaglin, F. Mosteller, and J. W. Tukey, New York: John Wiley, pp. 247–282.

Emerson, J. D. and Stoto, M. A. (1983), "Transforming Data," in *Understanding Robust and Exploratory Data Analysis*, eds. D. C. Hoaglin, F. Mosteller, and J. W. Tukey, New York: John Wiley, pp. 97–128.

Hoaglin, D. C. (1983), "Letter Values: A Set of Selected Order Statistics," in *Understanding Robust and Exploratory Data Analysis*, eds. D. C. Hoaglin, F. Mosteller, and J. W. Tukey, New York: John Wiley, pp. 33–57.

—— (1988), "Using Leverage and Influence to Introduce Regression Diagnostics," *College Mathematics Journal*, **19**, 387–401.

Hoaglin, D. C., Mosteller, F., and Tukey, J. W. (1983), *Understanding Robust and Exploratory Data Analysis*, New York: John Wiley.

Hoaglin, D. C. and Tukey, J. W. (1985), "Checking the Shape of Discrete Distributions," in *Exploring Data Tables, Trends, and Shapes*, eds. D. C. Hoaglin, F. Mosteller, and J. W. Tukey, New York: John Wiley, pp. 345–416.

Hoaglin, D. C. and Welsch, R. E. (1978), "The Hat Matrix in Regression and ANOVA," *The American Statistician*, **32**, 17–22, 146 (correction).

Huber, P. J. (1981), *Robust Statistics*, New York: John Wiley.

Iglewicz, B. (1983), "Robust Scale Estimators and Confidence Intervals for Location," in *Understanding Robust and Exploratory Data Analysis*, eds. D. C. Hoaglin, F. Mosteller, and J. W. Tukey, New York: John Wiley, pp. 404–431.

Li, G. (1985), "Robust Regression," in *Exploring Data Tables, Trends, and Shapes*, eds. D. C. Hoaglin, F. Mosteller, and J. W. Tukey, New York: John Wiley, pp. 281–343.

Mosteller, F. and Tukey, J. W. (1977), *Data Analysis and Regression*, Reading, MA: Addison-Wesley.

Rousseeuw, P. J. (1984), "Least Median of Squares Regression," *Journal of the American Statistical Association*, **79**, 871–880.

Rousseeuw, P. J. and Leroy, A. M. (1987), *Robust Regression and Outlier Detection*, New York: John Wiley.

Tukey, J. W. (1977), *Exploratory Data Analysis*, Reading, MA: Addison-Wesley.

Velleman, P. F. and Welsch, R. E. (1981), "Efficient Computing of Regression Diagnostics," *The American Statistician*, **35**, 234–242.

Wilk, M. B. and Gnanadesikan, R. (1968), "Probability Plotting Methods for the Analysis of Data," *Biometrika*, **55**, 1–17.

Chapter 9

Resistant and Robust Procedures

Thomas P. Hettmansperger
Pennsylvania State University

Simon J. Sheather
University of New South Wales

1. INTRODUCTION

Traditional courses in mathematical statistics, both elementary and advanced, include a discussion of likelihood-based inference methods. This highly successful conception is due mainly to Sir R. A. Fisher, who developed it in a series of papers in the 1920s; the volume *R. A. Fisher: An Appreciation* (Fienberg and Hinkley 1980) gives an account of this history. To recall and illustrate the basic ideas, we consider the simple location-scale model. Throughout we concentrate, especially in examples, on an assumed underlying normal model. Recall that the normal likelihood methods are identical to least-squares methods. Hence what we have to say applies also to the general approach of least squares.

It has long been known that inferences based on least-squares methods, constructed from the sample mean and standard deviation, are sensitive to assumptions about the underlying model. This was pointed out in an early paper by E. S. Pearson (1931). Surprisingly, the mean, the most efficient estimate at the normal model, can lose its optimality rapidly as the true model departs from the normal model. Tukey (1960) did pioneering work in this area. The significance level and confidence coefficient for the student t test and t confidence interval, constructed from the mean and standard deviation, are not particularly sensitive to the shape of the underlying distribution, as long as it is not highly skewed; however, the power of the test and the length of the interval can deteriorate quite rapidly when the underlying true model is close to but not precisely the normal model. We say

that a statistical inference procedure is *robust* when it is not sensitive to departures from the assumed model; that is, the efficiency, power, and significance level do not change dramatically as the model changes. The t test has robust significance level but not robust power. For this reason, we would say the t test is not robust. Robustness has the modest, but crucial, aim of retaining the assumed classical model while assuming in addition that the true underlying model is in a neighborhood of the assumed model. The implications of (slightly) differing assumed and true models are explored throughout this chapter. By taking this difference into account one obtains new, robust methods from the likelihood methods for the assumed model.

A statistical inference procedure is *robust* when it is not sensitive to departures from the assumed model; that is, the efficiency, power, and significance level do not change dramatically as the model changes. To be *resistant* a statistic, whether test or estimate, should not be controlled by a small portion of the data.

In addition to robustness properties of tests and estimates, resistance properties are also of interest. To be *resistant*, a statistic, whether test or estimate, should not be controlled by a small portion of the data. If a test statistic is not resistant, then a small amount of data contamination can determine whether the test rejects the null hypothesis. In the case of a nonresistant estimate, a few outliers are sufficient to push the value of the estimate to any value, or the introduction of a single outlier can cause a large change in the value of the estimate.

Our discussions quantify resistance in two ways. First we introduce influence functions. The *influence function* of an estimate measures the change in an estimate when an outlier is introduced. Estimates with bounded influence functions are desirable. Second, the *breakdown point* of an estimate is the fraction of observations that must be contaminated in order to force the estimate beyond any bound. A breakdown point of 50% is the best we can achieve. Similar concepts apply to tests, but we do not pursue them in this chapter. The mean and standard deviation are highly nonresistant estimates. One observation suffices to move them beyond any bound. The median, on the other hand, requires at least 50% contamination to move it beyond any bound. Further, the mean has an unbounded influence function; one outlying observation can induce an unbounded change in the mean. The median has bounded influence.

The ideas of robustness and resistance go hand in hand. Robustness is a property of the statistic as an inference procedure, whereas resistance is a property of the statistic as a function of the data. The strategy employed to robustify statistical procedures often produces resistant statistical methods as well. We first outline the large-sample theory of classical likelihood procedures, discuss both the resistance and robustness of these procedures, and introduce the new procedures based on a modification of the likelihood equations that offers hope of more resistant and robust behavior. We then devote much of the remainder of the chapter to showing that the new methods are indeed resistant and robust in the location-scale model. The regression setting introduces some new ideas on resistance, so we look at how our approach applies to simple linear regression. Finally, we indicate how these ideas extend to the multiple regression model.

2. BASIC LIKELIHOOD THEORY FOR THE LOCATION-SCALE MODEL

Let us begin with $Y_1, \ldots, Y_n$ independent and identically distributed (iid) random variables with common cumulative distribution function (cdf) $F((y - \theta)/\sigma)$, where $F(t)$ is a specified cdf with differentiable density func-

tion $f(t)$, $-\infty < \theta < \infty$ is an unknown location parameter, and $\sigma > 0$ is an unknown scale parameter. The *likelihood function* is given by:

$$
\begin{aligned}
L(\theta, \sigma) &= L(\theta, \sigma, y_1, \ldots, y_n) \\
&= \prod_{i=1}^{n} \frac{1}{\sigma} f\left(\frac{y_i - \theta}{\sigma}\right)
\end{aligned}
$$

where the lower-case $y_1, \ldots, y_n$ represent a realization of $Y_1, \ldots, Y_n$. To prepare for maximizing the likelihood as a function of θ and σ, we first take the natural logarithm and then differentiate to produce the likelihood equations. Let

$$
\begin{aligned}
\ell(\theta, \sigma) &= \ln L(\theta, \sigma) \\
&= \sum_{i=1}^{n} \ln f\left(\frac{y_i - \theta}{\sigma}\right) - n \ln \sigma
\end{aligned}
$$

be the "log-likelihood." The *maximum-likelihood estimate* (MLE) of (θ, σ) is the solution $(\hat{\theta}, \hat{\sigma})$ of the likelihood equations:

$$
\begin{aligned}
\frac{\partial}{\partial \theta} \ell(\theta, \sigma) &= 0 \\
\frac{\partial}{\partial \sigma} \ell(\theta, \sigma) &= 0.
\end{aligned}
$$

The canonical illustration takes

$$
f(t) = (2\pi)^{-1/2} \exp\{-t^2/2\},
$$

the normal model. Hence,

$$
\ell(\theta, \sigma) = -(2\sigma^2)^{-1} \Sigma (y_i - \theta)^2 - (n/2) \ln(2\pi\sigma^2)
$$

and we see the least-squares criterion appearing in the log-likelihood for the normal model. The likelihood equations become:

$$
\begin{aligned}
\frac{\partial}{\partial \theta} \ell(\theta, \sigma) &= \frac{1}{\sigma^2} \sum_{i=1}^{n} (y_i - \theta) = 0 \\
\frac{\partial}{\partial \sigma} \ell(\theta, \sigma) &= -\frac{n}{\sigma} + \frac{1}{\sigma^3} \sum_{i=1}^{n} (y_i - \theta)^2 = 0
\end{aligned}
$$

with the solutions (maximum-likelihood estimates)

$$
\hat{\theta} = \bar{y} \quad \text{and} \quad \hat{\sigma}^2 = n^{-1} \sum_{i=1}^{n} (y_i - \bar{y})^2
$$

or $\hat{\sigma} = (\hat{\sigma}^2)^{1/2}$. Note that $\hat{\sigma}^2$ is often corrected by replacing n^{-1} with $(n-1)^{-1}$ to produce an unbiased estimate of σ^2.

Patient	I (Hyoscyamine)	II (L-Hyoscine)	II−I
1	+0.7	+1.9	+1.2
2	−1.6	+0.8	+2.4
3	−0.2	+1.1	+1.3
4	−1.2	+0.1	+1.3
5	−0.1	−0.1	+0.0
6	+3.4	+4.4	+1.0
7	+3.7	+5.5	+1.8
8	+0.8	+1.6	+0.8
9	+0.0	+4.6	+4.6
10	+2.0	+3.4	+1.4

Table 1: Average sleep gain (hours per night) for patients given two drugs.

Example 1. Table 1 gives the average number (over several nights) of hours of sleep that each of ten patients gained from the use of two drugs. These data, published by Cushny and Peebles (1905), have often served as an example of "normal" data. The first and most famous use came when W.S. Gossett, writing under the pseudonym Student (1908), published his derivation of the t distribution along with an illustration of his t test. The normal quantile plot, shown in Figure 1, suggests that the underlying model may have tails heavier than the normal model and that an outlier may be present.

We wish to estimate θ, the center of the population of differences between the two drugs. The y-values are $1.2, 2.4, \ldots, 1.4$. If we assume a normal distribution for the differences, then the MLEs for θ and σ are $\hat{\theta} = \bar{y} = 1.58$ hours and $\hat{\sigma} = 1.16$ hours. From $\hat{\sigma}$ we estimate the standard error of $\hat{\theta}$ as $\hat{\sigma}/n^{1/2} = .37$. Thus, $\hat{\theta}$ is more than four estimated standard errors above 0, and this suggests that drug II is more effective.

The Location Model

As a first step toward nonnormal distributions and robustness, we focus on the pure location model. That is, we suppose, for the time being, that $\sigma = 1$. For a general density f, the likelihood equation becomes

$$
\begin{aligned}
\frac{d}{d\theta}\,\ell(\theta) &= \ell'(\theta) \\
&= -\sum_{i=1}^{n} \frac{f'(y_i - \theta)}{f(y_i - \theta)} = 0.
\end{aligned}
$$

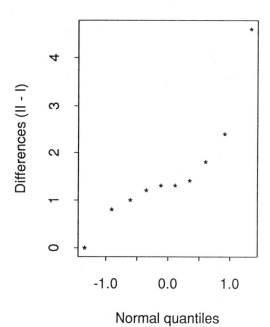

Figure 1: Normal quantile plot for the data of Example 1. The plot suggests nonnormality.

Anticipating robustness notation, we let

$$
\psi(t) = -\frac{f'(t)}{f(t)},
$$

and the likelihood equation becomes

$$
\ell'(\theta) = \sum_{i=1}^{n} \psi(y_i - \theta) = 0.
$$

As before, the solution $\hat{\theta}$ maximizes the likelihood of the observed realization $y_1, \ldots, y_n$.

In a similar development, we can construct a test of the null hypothesis $H_0 : \theta = 0$ versus $H_A : \theta \neq 0$. The rationale is simple: if $|\ell'(0)|$ is far from 0, then we reject $\theta = 0$ as a likely value. We must decide how large $|\ell'(0)|$ should be before we reject the null hypothesis. This is determined by our choice of significance level. Recall that the significance level α is the probability of committing a type I error; that is, rejecting the null hypothesis when the null hypothesis $H_0 : \theta = 0$ is true. Typical choices of α are .05 and .01. Hence, we seek a critical value c such that $P_0\left(|\ell'(0, Y_1, \ldots, Y_n)| > c\right) = \alpha$, the preassigned significance level, where P_0 means that the probability is computed under the null hypothesis, and we have included $Y_1, \ldots, Y_n$ in the notation for $\ell'(\cdot)$ to show that we now have a random quantity. When $|\ell'(0)| > c$, we reject the null hypothesis $H_0 : \theta = 0$ at level α and claim support for the alternative hypothesis. The statistic $\ell'(0) = \Sigma\psi(y_i)$ is called the *likelihood scores statistic*.

It may be difficult to derive the distribution of $\ell'(0, Y_1, \ldots, Y_n) = \Sigma\psi(Y_i)$ under $H_0 : \theta = 0$. However, if $E_0\psi^2(Y) < \infty$ and $E_0\psi(Y) = 0$, the central limit theorem allows us to approximate the critical value as

$$\frac{c}{(nE_0\psi^2(Y))^{1/2}} \doteq z_{\alpha/2}$$

where $1 - \Phi(z_{\alpha/2}) = \alpha/2$ and $\Phi(\cdot)$ denotes the standard normal cdf. For example, if $\alpha = .05$, then $c \doteq 1.96\,(nE_0\psi^2(Y))^{1/2}$, and the approximate size .05 test of $H_0 : \theta = 0$ vs. $H_A : \theta \neq 0$ rejects H_0 if

$$\left|\sum_{i=1}^{n} \psi(y_i)\right| > 1.96\,(nE_0\psi^2(Y))^{1/2}.$$

An alternative approach to the likelihood scores test works with $\hat{\theta}$ directly. If $n^{1/2}\hat{\theta}$ is asymptotically normally distributed with mean 0 and variance $\sigma_{\hat{\theta}}^2$ under $H_0 : \theta = 0$, then an approximate size α test rejects $H_0 : \theta = 0$ in favor of $H_A : \theta \neq 0$ when $|\hat{\theta}| > z_{\alpha/2}\sigma_{\hat{\theta}}$. We use this alternative approach in our examples; it has the same large-sample (asymptotic) properties as the score test.

In Example 1, with $\sigma \equiv 1$, approximation is not necessary if we assume a normal underlying distribution. We have $\psi(t) = t$, $\ell'(0) = \Sigma y_i$ and when $H_0 : \theta = 0$ is true, ΣY_i has a normal distribution with mean 0 and variance n. Hence, the exact level .05 test rejects $H_0 : \theta = 0$ when $|\Sigma y_i| > 1.96\,\sqrt{n}$ since $E_0\psi^2(Y) = E_0 Y^2 = \sigma^2 \equiv 1$. But, this is the same as the test based on $\hat{\theta}$ since $\hat{\theta} = \Sigma y_i/n$.

In a traditional course in mathematical statistics, likelihood methods would be connected to the ideas of sufficiency and finite-sample efficiency via the exponential family of underlying distributions; see Arnold (1990, Section 10.2.2). In this chapter we consider, instead, the asymptotic efficiency and power properties of the maximum-likelihood estimate $\hat{\theta}$ and the likelihood scores test based on $\ell'(0) = \Sigma\psi(y_i)$, respectively. Further, we introduce and discuss robustness and resistance properties of the estimator.

The discussion of these properties of the likelihood meth-

ods will provide motivation for the introduction of robust methods. The robust methods are modifications of the likelihood methods and have been dubbed M-estimates and M-tests to reflect their origins in maximum likelihood theory by Huber (1964), who laid the foundations for this area.

Good reviews of the traditional likelihood methods are provided by Arnold (1990) and Bickel and Doksum (1977). Books on robust methods include the following: at an intermediate level Staudte and Sheather (1990), and at a more advanced level Huber (1981) and Hampel et al. (1986). A few additional references will be given throughout the discussion, along with specific references to these monographs. Generally, however, the reader can find further discussion of most topics in at least one of these books.

3. BASIC LIKELIHOOD THEORY

We begin with the estimating equation

$$\Sigma\psi(y_i - \theta) = 0$$

where the basic example will initially be the likelihood score $\psi(t) = -f'(t)/f(t)$. We then move on to a robust modification of ψ. In addition, we consider the scores test statistic $\Sigma\psi(y_i)$ for testing $H_0 : \theta = 0$.

The development is heuristic, and we suppose that differentiation and integration can be interchanged. We assume that the function ψ is nondecreasing, crosses the horizontal axis, and is sufficiently smooth to allow differentiation (with the possible exception of a finite set of points). For a rigorous development with regularity conditions see Huber (1981, Chapter 3).

We write $E\psi^2$ for $E\psi^2(Y) = \int \psi^2(y)f(y)dy$ and $E\psi'$ for $E\psi'(Y) = \int \psi'(y)f(y)dy$; and we suppose that $E\psi(Y) = 0$, $E\psi' > 0$, and $0 < E\psi^2 < \infty$.

We finally suppose $f(t)$ is symmetric about 0, so that θ can be identified as the center of symmetry of the underlying distribution. The asymptotic properties of the estimate and test can be deduced from a simple linear approximation to the likelihood function. Essentially, the linear approximation is like an expansion in θ of the estimating equation about the true value of θ. The solution $\hat{\theta}$ of the estimating equation has the same limiting properties as the solution $\tilde{\theta}$ of the linear approximation, which then can be handled quite easily.

Estimation

Proposition 1: *If the true value of θ is 0, then*

$$n^{-1/2} \sum_{i=1}^{n} \psi(Y_i - \theta) = n^{-1/2} \sum_{i=1}^{n} \psi(Y_i) - n^{1/2}\theta E\psi' + o_p(1)$$

where as n increases $o_p(1)$ converges to zero in probability, uniformly for all θ such that $n^{1/2}|\theta| \leq K$, for any $K > 0$.

A simple argument can be given for the case when $|\psi''(t)|$ is bounded by a constant B. Expand in θ about 0 to get

$$\begin{aligned} n^{-1/2} \sum_{i=1}^{n} \psi(Y_i - \theta) &= n^{-1/2} \sum_{i=1}^{n} \psi(Y_i) \\ &\quad - n^{1/2}\theta \left(n^{-1} \sum_{i=1}^{n} \psi'(Y_i)\right) + R_n. \end{aligned}$$

Then $n^{-1}\Sigma\psi'(Y_i)$ converges in probability to $E\psi'$ by the Weak Law of Large Numbers. Further, because R_n contains the bounded second derivative of ψ,

$$|R_n| \leq |n\theta^2 B n^{-1/2}| \leq K^2 B n^{-1/2}$$

so $R_n = o_p(1)$, uniformly for $n^{-1/2}|\theta| \leq K$. Hence, the proposition follows.

Defining $S(\theta) = \Sigma\psi(Y_i - \theta)$, we write the result in Proposition 1 as $n^{-1/2}S(\theta) \doteq n^{-1/2}S(0) - n^{1/2}\theta E\psi'$. We are now ready to state the approximating distribution for $n^{1/2}\hat{\theta}$ when $\theta = 0$ is the true parameter value.

Proposition 2: *Let $\hat{\theta}$ be the solution of the equation $\Sigma\psi(y_i - \theta) = 0$. If $\theta = 0$ is the true parameter value, then $n^{1/2}\hat{\theta}$ converges in distribution to a random variable which is normal with mean 0 and variance $E\psi^2/(E\psi')^2$.*

More compactly, $n^{1/2}\hat{\theta} \overset{\mathcal{D}}{\to} N(0, E\psi^2/(E\psi')^2)$, where $N(a, b^2)$ denotes a normal distribution with mean a and variance b^2. We might also say that $\hat{\theta}$ is approximately normal with mean 0 and variance $E\psi^2/n(E\psi')^2$.

This result can be anticipated from the result in Proposition 1. The solution $\hat{\theta}$ of the equation $\Sigma\psi(y_i - \theta) = 0$ behaves like the solution $\tilde{\theta}$ to the approximation $n^{-1/2}\Sigma\psi(y_i) - n^{1/2}\tilde{\theta}E\psi' = 0$, because the remainder is $o_p(1)$. Now

$$n^{1/2}\tilde{\theta} = \frac{1}{E\psi'} n^{-1/2}\Sigma\psi(Y_i)$$

and the central limit theorem implies that $n^{-1/2}\Sigma\psi(Y_i)$ converges in distribution to $N(0, E\psi^2)$. Hence $n^{1/2}\tilde{\theta}$ (and $n^{1/2}\hat{\theta}$) converges in distribution to $N(0, E\psi^2/(E\psi')^2)$.

Testing

Next consider the test of $H_0 : \theta = 0$ vs. $H_A : \theta > 0$. Reject $H_0 : \theta = 0$, at approximate significance level α, when $S = \Sigma\psi(y_i) > z_\alpha(nE\psi^2)^{1/2}$, where $1 - \Phi(z_\alpha) = \alpha$. The power of the test at the point $\theta^* > 0$ will approach 1 as n increases. This makes it difficult to compare the approximate powers of such tests for large sample sizes and fixed significance levels, because they all have roughly the same power. To circumvent this difficulty, we consider the power along a sequence of alternatives $n^{-1/2}\theta^*$ converging to 0. By choosing the sequence $n^{-1/2}\theta^*$, we offset the effects of increasing sample size by shrinking the alternative at just the right rate toward the null hypothesis. The power will converge to a value less than one, and so size α tests can be compared along this sequence. The limit of the power function along the sequence $n^{-1/2}\theta^*$ is called the *asymptotic* (for large n) *local* (because the alternative is close to 0) *power* of the test. The next proposition states this limit.

Proposition 3: *Consider testing $H_0 : \theta = 0$ versus $H_A : \theta_n = n^{-1/2}\theta^*$ with $\theta^* > 0$ fixed. The asymptotic local power of the size α test that rejects $H_0 : \theta = 0$ if $S = \Sigma\psi(y_i) > z_\alpha(nE\psi^2)^{1/2}$ is given by*

$$\lim_{n \to \infty} P_{\theta_n}(S > z_\alpha(nE\psi^2)^{1/2}) =$$
$$1 - \Phi(z_\alpha - \theta^* E\psi'/(E\psi^2)^{1/2}).$$

Again, the result is easy to anticipate from Proposition 1. Write $\tilde{S}(\theta) = \Sigma\psi(Y_i - \theta)$, so that $S = S(0)$. Now, because we have a location model,

$$P_{\theta_n}(S(0) > z_\alpha(nE\psi^2)^{1/2}) = P_0(S(-\theta_n) > z_\alpha(nE\psi^2)^{1/2})$$

and we may use $n^{-1/2}S(-\theta_n) \doteq n^{-1/2}S(0) + n^{1/2}\theta_n E\psi'$. Hence the power is approximated by:

$$\begin{aligned} P_0(S(0) &> z_\alpha(nE\psi^2)^{1/2} - n\theta_n E\psi') \\ &= P_0(n^{-1/2}S(0)/(E\psi^2)^{1/2} \\ &> z_\alpha - \theta^* E\psi'/(E\psi^2)^{1/2}). \end{aligned}$$

The result now follows from the central limit theorem because $n^{-1/2}S(0)/(E\psi^2)^{1/2}$ converges in distribution to $N(0, 1)$.

Efficacy

We define the *efficacy factor* e_ψ by

$$e_\psi = \frac{E\psi'}{(E\psi^2)^{1/2}}.$$

The efficacy factor measures the performance of the estimate or test in terms of the asymptotic variance and the asymptotic local power, respectively. The factor is a standardized expected rate of change of ψ at the true parameter value $\theta = 0$. A large value of e_ψ means that ψ responds rapidly to changes in θ near the true value. Hence, the scores test should be sensitive to changes from the null-hypothesis value of $\theta = 0$. Further, $n^{1/2}\hat\theta$ is approximately $N(0, 1/e_\psi^2)$, and the approximate power of $S = \Sigma\psi(y_i)$ along $\theta_n = n^{-1/2}\theta^*$, $\theta^* > 0$, is $1 - \Phi(z_\alpha - \theta^* e_\psi)$. Hence, large values of e_ψ mean small asymptotic variance and large asymptotic local power.

Recall that f specifies the assumed underlying model, so that $\psi(t) = -f'(t)/f(t)$. Supposing this assumption is correct, we have

$$E\psi^2 = E\left[\frac{f'(Y)}{f(Y)}\right]^2$$

$$E\psi' = -E\left[\frac{f''(Y)f(Y) - (f'(Y))^2}{(f(Y))^2}\right].$$

But if expectation and differentiation can be interchanged, then $E\psi^2 = E\psi'$, and both are equal to the *Fisher information*, denoted by I_f. Hence $e_\psi = I_f^{1/2}$; $n^{1/2}\hat\theta$, the MLE, is approximately $N(0, 1/I_f)$; and the asymptotic local power is $1 - \Phi(z_\alpha - \theta^* I_f^{1/2})$. The Cramér-Rao lower bound, discussed in a basic mathematical statistics course, suggests that $n^{1/2}\hat\theta$ is asymptotically efficient; that is, it has the smallest asymptotic variance, and also the scores test has the maximum asymptotic local power. See Arnold (1990, Sections 7.2.3 and 7.3.1)

We now ask what happens when the true distribution differs from the assumed distribution. This question provides the motivation to seek a robustification of the likelihood methods.

Suppose we assume the underlying distribution is f but the true distribution is g. Let ψ_f denote $-f'(t)/f(t)$. Then we must compute

$$E_g\psi'_f = \int \psi'_f(t)g(t)dt$$

$$E_g\psi_f^2 = \int \psi_f^2(t)g(t)dt$$

to determine the efficacy factor

$$e(\psi_f|g) = E_g\psi'_f/(E_g\psi_f^2)^{1/2}$$

where we have expanded the notation to distinguish between the assumed and true models. The appropriate score is, of course, ψ_g. The next proposition shows that we indeed lose efficiency and power as suggested.

Proposition 4:   *If g is the true model and f is the assumed model, then $e(\psi_g|g) \geq e(\psi_f|g)$.*

The argument is an application of the Cauchy-Schwarz inequality similar to the argument for the Cramér-Rao lower bound.

$$E_g\psi'_f = \int \psi'_f(t)\,g(t)\,dt$$

$$= -\int \psi_f(t)g'(t)\,dt \quad \text{(integrating by parts)}$$

$$= \int \psi_f(t)g^{1/2}(t)\left(-\frac{g'(t)}{g^{1/2}(t)}\right)\,dt$$

$$\leq \left\{\int \psi_f^2(t)g(t)\,dt \cdot \int\left(\frac{g'(t)}{g(t)}\right)^2 g(t)\,dt\right\}^{1/2}$$

$$= (E_g\psi_f^2)^{1/2}\, I_g^{1/2}$$

$$e(\psi_f|g) = E_g\psi'_f/(E_g\psi_f^2)^{1/2} \leq I_g^{1/2} = e(\psi_g|g).$$

We suppose $E_f\psi_g = \int \psi_g(t)f(t)\,dt = 0$ so $\psi_g(t)f(t) \to 0$ as $t \to \pm\infty$, used in the preceding integration by parts. The proposition says two things: (a) if we are correct and the assumed f is correct, then over the class of methods generated by different likelihood scores the correct likelihood methods are asymptotically optimal; and (b) if we are wrong, then there is a better score function, namely ψ_g where g is the true, but unknown, distribution.

4. ROBUST AND RESISTANT METHODS

We now face the following question: How much do we lose by not knowing the true model? From Proposition 4, we know that we must lose something. We show in Example 2 below that, when we assume a normal model and use the corresponding likelihood methods that are optimal for this model, we may lose considerable efficiency when the true model differs only slightly from the assumed model. We give up the hope of using an

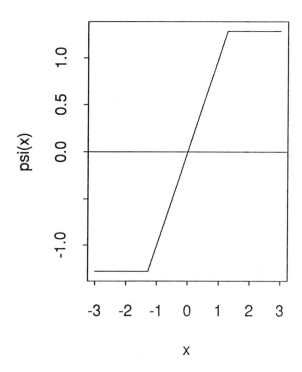

Figure 2: Huber's ψ function.

a truncated version of ψ_f:

$$
\begin{aligned}
\psi(t) &= \max\{-k, \min[\psi_f(t), k]\} \\
&= \min\{k, \max[\psi_f(t), -k]\}
\end{aligned}
$$

for some constant k. In effect, ψ behaves like ψ_f for the central portion of the data and then diminishes the influence of extreme data on the determination of $\hat\theta$.

More generally, any $\hat\theta$ that satisfies $\Sigma\psi(y_i - \hat\theta)$ for some suitable choice of ψ is called an *M-estimate* because it is motivated by maximum likelihood.

Example 2. Suppose we assume a normal model as the center of the neighborhood. Then $\psi_f(t) = t$ and $\psi(t) = \max[-k, \min(t, k)]$. Other representations are

$$
\begin{aligned}
\psi(t) &= \min\left\{1, \frac{k}{|t|}\right\} t \\
&= \begin{cases} k\,\mathrm{sgn}\,t & |t| \ge k \\ t & |t| \le k \end{cases}
\end{aligned}
$$

This ψ, called *Huber's ψ function*, is the focus of our attention for much of the remainder of the chapter. Figure 2 shows the graph of ψ.

The structure of the neighborhood determines the best value of k in Huber's ψ. We next compare normal likelihood methods to methods generated from ψ. The idea is to find some workable value of k that provides a high efficacy at the assumed normal model and remains high throughout the neighborhood.

optimal procedure and instead search for a method that works well everywhere in a neighborhood of the assumed model. Then, if the true model is in this neighborhood, we have a good but not optimal procedure. Such a procedure is robust in the sense that it is not overly sensitive to the model assumption. The efficiency loss is the insurance premium paid to cover not knowing the true model.

Huber's Approach

As a general strategy, Huber (1964) supposed the true density g is in a neighborhood of the assumed density f. Let h be a possible model in the neighborhood of f. Then an overall measure of the quality of the test and estimate based on ψ_h is $\min\{e(\psi_h|g): g$ in the neighborhood$\}$. We can compare minimum efficacy factors for various distributions h to see whether there is a best one. Indeed, Huber showed that, under regularity conditions, there is a score function ψ that maximizes the minimum efficacy over the neighborhood. This ψ function is simply the likelihood score function corresponding to the distribution with minimum Fisher information in the neighborhood. Hence, the strategy is to do the best you can in the least favorable situation. The resulting ψ function is

---

How much do we lose by not knowing the true model? From Proposition 4, we know that we must lose something.

Our general model is now the following: We specify a normal model with mean θ and variance 1 (later we reintroduce the unknown variance σ^2) at the center of the neighborhood. The true model, denoted by g, is a symmetric distribution centered at θ and located in the neighborhood. With a finite sample, it is difficult, if not impossible, to statistically distinguish the assumed model from the true model. We now have two sets of methods:

(a) Normal likelihood methods (which correspond to the method of least squares): $\bar y$ as an estimate of θ and Σy_i as the basis of the test of $H_0 : \theta = 0$.

k	ϵ:	0	$\tau = 3$ .05	.10	$\tau = 5$ .05	.10
1		0.903	1.153	1.350	1.772	2.439
1.1		0.919	1.169	1.365	1.795	2.458
1.2		0.933	1.183	1.376	1.812	2.470
1.28		0.943	1.191	1.382	1.823	2.475
1.3		0.945	1.193	1.384	1.825	2.475
1.4		0.955	1.201	1.388	1.834	2.473
1.5		0.964	1.207	1.389	1.838	2.465
1.6		0.972	1.210	1.387	1.839	2.452
1.7		0.978	1.212	1.383	1.836	2.433
1.8		0.983	1.212	1.377	1.831	2.410
1.9		0.987	1.210	1.369	1.823	2.383
2.0		0.990	1.207	1.360	1.812	2.353

Table 2: Efficiency of M-estimate $\hat{\theta}$ relative to $\bar{y}$ for normal and contaminated normal models.

(b) Truncated normal likelihood methods (from Example 2): $\hat{\theta}$ such that $\Sigma\psi(y_i - \hat{\theta}) = 0$ as the estimate and $\Sigma\psi(y_i)$ as the basis of the test, where ψ is Huber's ψ function.

We wish to compare the properties of the two methods (a) and (b) at the assumed model and at various possible true models, g.

For (a), $EY = 0$ and $EY^2 = \int t^2 g(t)\, dt$. The efficacy factor is the reciprocal of the standard deviation of $n^{1/2}\bar{y}$, namely $e = 1/(EY^2)^{1/2}$. When g is the standard normal density, $e = 1$.

For (b), $E\psi^2 = 2k^2 G(-k) + \int_{-k}^{k} t^2 g(t)\, dt$, $E\psi' = 1 - 2G(-k)$, and the efficacy factor is

$$e_\psi = \frac{1 - 2G(-k)}{\{2k^2 G(-k) + \int_{-k}^{k} t^2 g(t)\, dt\}^{1/2}}.$$

The efficiency of methods (b), both estimate and test, relative to methods (a) is defined to be eff $= e_\psi^2/e^2$. Hence, if eff < 1, then methods (a) have smaller asymptotic variance and larger asymptotic local power and are superior, whereas if eff > 1, then methods (b) are superior.

To illustrate the efficiency, we consider the *contaminated normal distribution*, which has density

$$g(t) = (1 - \epsilon) f(t) + \epsilon\tau^{-1} f(\tau^{-1} t)$$

for $0 \le \epsilon \le 1$, $\tau > 0$, and $f(t)$ the standard normal density function. This distribution was introduced and discussed by Tukey (1960). It consists of a mixture of the standard normal distribution and another normal distribution with standard deviation τ. By varying ϵ and τ we generate a neighborhood of distributions with the standard normal distribution ($\epsilon = 0$) at the center. It is very difficult to detect the difference between an assumed normal model and a true contaminated normal model. Many studies take $\tau = 3$. This choice is partly motivated by the fact that, when $\epsilon = 0.1$ and $\tau = 3$, the assumed standard normal distribution and the contaminating component contribute equal amounts to the overall variance of the true model.

Table 2 gives the efficiencies for $\epsilon = 0$, .05, 0.1, for $\tau = 3$ and 5, and k between 1 and 2. When the true model is normal, the efficiency is greater than 90 percent. When the model is contaminated, the Huber M-estimate is always more efficient for the chosen example. For example, if the true model is a 10% contaminated normal model with $\tau = 3$, then the Huber M-estimate $\hat{\theta}$ with $k = 1.3$ is 38% more efficient than the normal maximum-likelihood estimate $\bar{y}$. A value of k around 1.3 works quite well for the contamination neighborhood. We use 1.28, which is recommended by Staudte and Sheather (1990, p. 132). Huber (1964, p. 82) suggests any value of k between 1 and 2, based on analyzing fairly rich neighborhoods. Hampel et al. (1986, Section 1.2) point out that data contamination in practice appears to be more the rule than the exception.

Computing the Estimates

Now that we have a recommended value for k, the next important issue is how to actually compute $\hat{\theta}$, the solution of $\Sigma \psi (y_i - \theta) = 0$. In practice, we also have to scale the data, because we cannot reasonably expect the variance σ^2 to be one. We use the median absolute deviation from the median (MAD) and define

$$\hat{\sigma} = 1.483 \ \text{med}_i | \, y_i - \text{med}_j y_j |,$$

see Hampel et al. (1986, p. 105). The multiplicative factor makes $\hat{\sigma}$ a consistent estimate of the standard deviation of the assumed normal model. Hence, we actually solve

$$\sum_{i=1}^{n} \psi \left(\frac{y_i - \theta}{\hat{\sigma}} \right) = 0.$$

From the alternative representation in Example 2, we write this equation as

$$\Sigma w_i (\theta) \left(\frac{y_i - \theta}{\hat{\sigma}} \right) = 0$$

where

$$w_i (\theta) = \min \left\{ 1, \frac{k\hat{\sigma}}{|y_i - \theta|} \right\}.$$

This is called a *weighted least-squares formulation*. The weight $w_i(\theta)$ depends on the data as well as on θ; however, for simplicity, we show only the dependence on θ in the notation. The solution is arrived at iteratively:

1. Find an initial estimate $\hat{\theta}^{(0)}$, say $\bar{y}$ or med y_i.

2. Compute $\hat{\sigma}$ and $w_i (\hat{\theta}^{(0)})$.

3. Solve to find

$$
\begin{aligned}
\hat{\theta}^{(1)} &= \frac{\Sigma w_i (\hat{\theta}^{(0)}) \, y_i}{\Sigma w_i (\hat{\theta}^{(0)})} \\
&= \frac{\Sigma w_i (\hat{\theta}^{(0)})(y_i - \hat{\theta}^{(0)})}{\Sigma w_i (\hat{\theta}^{(0)})} + \hat{\theta}^{(0)}.
\end{aligned}
$$

4. Repeating steps 1–3 with $\hat{\theta}^{(0)}$ replaced by $\hat{\theta}^{(1)}$ produces $\hat{\theta}^{(2)}$; $\hat{\sigma}$ remains fixed.

5. Continue repeating the steps until $|\hat{\theta}^{(j+1)} - \hat{\theta}^{(j)}|$ is small, at which point stop and take $\hat{\theta} = \hat{\theta}^{(j)}$ as the solution.

This simple iterative computing method works quite well and converges rapidly to the solution. Note that the weight $w_i (\theta) = \min\{1, k\hat{\sigma}/| \, y_i - \theta|\}$ smoothly downweights data in proportion to how much $|y_i - \theta|$ exceeds $\hat{\sigma}$, with extreme data downweighted the most. If no observations are extreme, that is, $| \, y_i - \theta | \leq k\hat{\sigma}$ for $i = 1, \ldots, n$, then no observations are downweighted and $\hat{\theta} = \bar{y}$, the MLE of the assumed model. This smooth downweighting shows how the Huber ψ function robustifies the likelihood methods, and this will be the central strategy in more complex experimental designs to be discussed below.

A second iterative approach uses *Newton's method*. We briefly discuss it here because it is faster for the solution of the location problem, and we apply it to our example. Later we return to the weighted-least-squares formulation because it is more widely available on standard packages.

In the argument for Proposition 1, suppose we expand $\Sigma \psi (y_i - \theta)$ about an initial estimate $\hat{\theta}^{(0)}$ rather than about 0. Then we have, using $\hat{\sigma} = 1.483$ MAD,

$$
\begin{aligned}
n^{-1} \Sigma \psi \left(\frac{y_i - \theta}{\hat{\sigma}} \right) &\doteq n^{-1} \Sigma \psi \left(\frac{y_i - \hat{\theta}^{(0)}}{\hat{\sigma}} \right) \\
&- (\theta - \hat{\theta}^{(0)}) \left(n^{-1} \Sigma \frac{1}{\hat{\sigma}} \psi' \left(\frac{y_i - \hat{\theta}^{(0)}}{\hat{\sigma}} \right) \right).
\end{aligned}
$$

Setting this equal to 0 and solving yields

$$\hat{\theta}^{(1)} = \hat{\theta}^{(0)} + \hat{\sigma} \frac{\Sigma \psi [(y_i - \hat{\theta}^{(0)})/\hat{\sigma}]}{\Sigma \psi' [(y_i - \hat{\theta}^{(0)})/\hat{\sigma}]}.$$

This process can now be iterated until $|\hat{\theta}^{(j+1)} - \hat{\theta}^{(j)}|$ is small; see Staudte and Sheather (1990, p. 117).

Unknown Scale

For applications, we need to revise the conclusion of Proposition 2 to account for the unknown scale. Because we replace $\psi (y - \theta)$ by $\psi [(y - \theta)/\sigma]$, we must then replace $\psi'(y - \theta)$ by $\sigma^{-1}\psi'[(y - \theta)/\sigma]$. Then we have for Proposition 2 that $n^{1/2}(\hat{\theta} - \theta)$ converges in distribution to $N(0, \sigma^2 E\psi^2/(E\psi')^2)$. In other words, $\hat{\theta}$ is approximately normally distributed with mean 0 and variance $\sigma^2 E\psi^2/n(E\psi')^2$. Then σ^2 is replaced by $\hat{\sigma}^2 = 1.483$ MAD, $E\psi^2$ is replaced by $n^{-1} \Sigma \psi^2 [(y_i - \hat{\theta})/\hat{\sigma}]$ and $E\psi'$ by $n^{-1}\Sigma \psi'[(y_i - \hat{\theta})/\hat{\sigma}]$.

Example 3. We return to Example 1 and illustrate some of the computations. As pointed out, Newton's method yields an iterative solution beginning with an

i	y_i	$\frac{y_i - 1.3}{0.593}$	$\psi\left(\frac{y_i - 1.3}{0.593}\right)$	$\psi'\left(\frac{y_i - 1.3}{0.593}\right)$
1	1.2	−0.169	−0.169	1
2	2.4	1.855	1.280	0
3	1.3	0.000	0.000	1
4	1.3	0.000	0.000	1
5	0.0	−2.192	−1.280	0
6	1.0	−0.506	−0.506	1
7	1.8	0.843	0.843	1
8	0.8	−0.843	−0.843	1
9	4.6	5.565	1.280	0
10	1.4	0.169	0.169	1
			0.774	7

Table 3: One-step M-estimate for Example 3.

initial estimate $\hat{\theta}^{(0)}$. If we take only one step of the iteration, then $\hat{\theta}^{(1)}$ is called a *one-step M-estimate*.

We show how to compute $\hat{\theta}^{(1)}$ with $k = 1.28$ starting from $\hat{\theta}^{(0)} = \text{med } y_i = 1.30$ and $\hat{\sigma} = 1.483$ MAD $= .593$. Table 3 provides the needed information.

$$
\hat{\theta}^{(1)} = \hat{\theta}^{(0)} + \frac{\hat{\sigma} \sum_{i=1}^{n} \psi[(y_i - \hat{\theta}^{(0)})/\hat{\sigma}]}{\sum_{i=1}^{n} \psi'[(y_i - \hat{\theta}^{(0)})/\hat{\sigma}]}
$$
$$
= 1.3 + 0.593 \times 0.774/7 = 1.37.
$$

The estimate of the asymptotic variance

$$
\frac{\sigma^2 E\psi^2}{n(E\psi')^2}
$$

is .053. Hence, the estimated standard error of $\hat{\theta}^{(1)}$ is .23, and $\hat{\theta}^{(1)} = 1.37$ is almost 6 standard errors above 0. This suggests a stronger statement of significance than that based on the normal likelihood estimate in Example 1. The outlier 4.6 diminished the sensitivity of $\bar{y} = 1.58$ by inflating the estimate of the standard error $\hat{\sigma}/n^{1/2} = .37$; on the other hand, $\hat{\theta}^{(1)}$ and its standard error are more resistant. Figure 3 shows the outlier.

Bootstrap Estimate of Standard Error

We end this example with a brief discussion of the bootstrap method of estimating the standard error of $\hat{\theta}^{(1)}$. The idea is to simulate the standard error by repeated

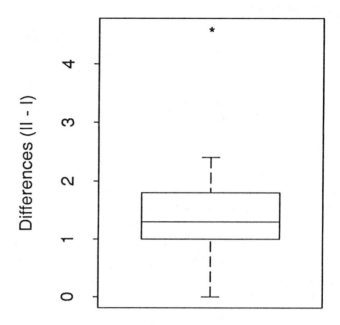

Figure 3: Boxplot of the data from Example 1.

sampling with replacement from the original data. We drew 200 samples of size 10 from the data and computed the values of $\hat{\theta}^{(1)}$ for each sample. The sample standard deviation of these values is the bootstrap estimate of the standard error of $\hat{\theta}^{(1)}$. Figure 4, modeled after a similar display in Diaconis and Efron (1983), displays the steps for constructing a bootstrap estimate. For the bootstrap computation, we used the Minitab macros in Staudte and Sheather (1990, appendix). We found this estimate to be .20. This is roughly the same as our estimated asymptotic standard error of .23. For further discussion of the bootstrap, see the highly readable account by Diaconis and Efron (1983) or Efron (1982). The bootstrap belongs to a class of resampling methods that has an extensive literature.

5. RESISTANCE PROPERTIES OF ESTIMATES

In addition to efficiency, the stability of a statistical method involves important resistance issues. We confine our attention to estimators and first discuss influence. For a technical definition of influence, see Staudte and Sheather (1990, Section 3.2.2). Our discussion is heuristic and based on the sample data.

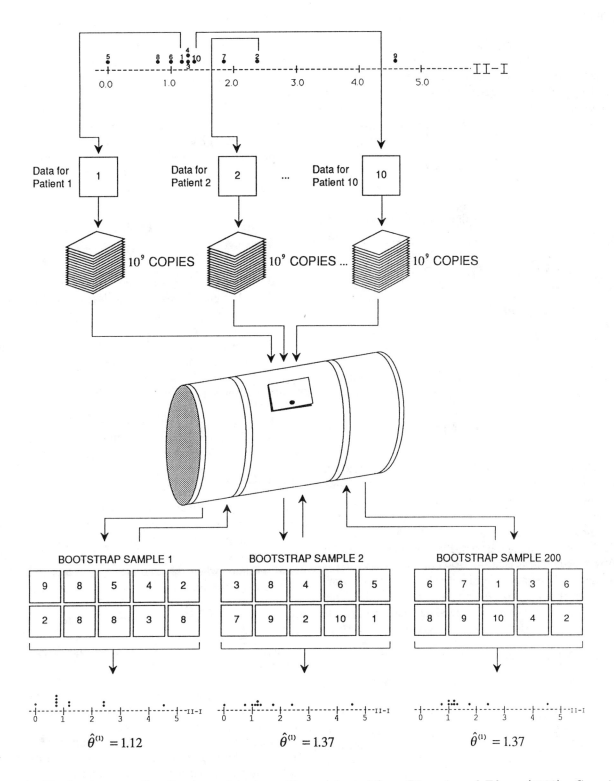

Figure 4: The bootstrap estimation procedure in outline. Adapted from Diaconis and Efron, (1983). Copyright ©1983 by Scientific American, Inc. All rights reserved.

Influence

For simplicity of presentation we once again take $\sigma = 1$. Also we attach a subscript to $\hat{\theta}$ to indicate sample size. Suppose $y_1, \ldots, y_n$ are n data values such that $\hat{\theta}_n = 0$, so $\Sigma \psi(y_i) = 0$. The 0 value is chosen for convenience. Suppose we add another data point $y_{n+1} = y^*$ and we wish to see how rapidly $\hat{\theta}$ changes. Let $\hat{\theta}_{n+1}$ solve

$$(n+1)^{-1/2} \sum_{i=1}^{n+1} \psi(y_i - \theta) = 0.$$

Because $\sum_1^n \psi(y_i) = 0$, the linear approximation from Proposition 1

$$(n+1)^{-1/2} \sum_1^{n+1} \psi(y_i) - (n+1)^{1/2} \hat{\theta}_{n+1} E\psi' \doteq 0$$

reduces to

$$(n+1)^{-1/2} \psi(y^*) - (n+1)^{1/2} \hat{\theta}_{n+1} E\psi' \doteq 0.$$

Hence $(n+1) \hat{\theta}_{n+1} \doteq \psi(y^*)/E\psi'$. This is an approximation to the derivative; and, if we reintroduce $\hat{\theta}_n$, we can write:

$$\frac{\hat{\theta}_{n+1} - \hat{\theta}_n}{\frac{1}{n+1} - 0} \doteq \frac{\psi(y^*)}{E\psi'}$$

Hence, let $\Omega(y^*) = \psi(y^*)/E\psi'$. The function $\Omega(y^*)$ is called the *influence function* of the estimate $\hat{\theta}$. It roughly measures the impact of an outlier on the estimate. An estimate is said to be *qualitatively robust* if the influence function is bounded, as it is for $\hat{\theta}$. This means the estimate cannot change arbitrarily fast when an outlier is introduced. Note that for $\bar{y}$, $\Omega(y^*) = y^*$, which is unbounded. Thus a single observation can have an arbitrarily large impact on $\bar{y}$, and so $\bar{y}$ is not qualitatively robust. The reader may wish to show that the sample median has a bounded influence function.

Finally, recall from the argument of Proposition 2 that

$$n^{1/2} \hat{\theta} \doteq n^{-1/2} \Sigma \psi(Y_i)/E\psi' = n^{-1/2} \Sigma \Omega(Y_i).$$

Hence, the influence function represents the first-order approximation for the estimate, and Proposition 2 could be restated to say that $n^{1/2} \hat{\theta}$ is asymptotically normal with mean 0 and variance $E\Omega^2(y)$, since $E\Omega(y) = 0$. For a rigorous approach to influence functions, see Huber (1981) or Hampel et al. (1986).

Breakdown Point

The other important aspect of resistance is the *breakdown point* of the estimator. This is the smallest fraction of the data that must be contaminated in order to drive the estimate beyond any bound. The breakdown point b^* for $\bar{y}$ is $b^* = n^{-1}$, because one arbitrary observation is sufficient to carry the mean away. Since b^* converges to 0, we say that $\bar{y}$ has breakdown point 0. The next proposition shows that M-estimates with $k > 0$ have the best possible breakdown point, just like the highly resistant sample median.

Proposition 5: *For all $k > 0$, the Huber M-estimate $\hat{\theta}$ has breakdown point .5.*

We present an argument for the case of odd sample size: $n = 2m+1$. Suppose we move the last $m+1$ observations to $+\infty$. Then, $\sum_1^{2m+1} \psi(y_i - \theta) = \sum_{i=1}^m \psi(y_i - \theta) + (m+1)k = 0$ is our equation. But $\sum_{i=1}^m \psi(y_i - \theta) \geq -mk$, so no value of θ solves the equation. On the other hand, if we move only m observations to $+\infty$, we have $\sum_{i=1}^{m+1} \psi(y_i - \theta) + mk = 0$, which always has a finite solution. Hence $b^* = (m+1)/n$, which converges to .5.

Note that $\hat{\sigma} = 1.483$ MAD also has breakdown point .5, so that using $\hat{\sigma}$ to scale the data does not affect the result in Proposition 5. Hence, $\hat{\theta}$ has breakdown .5 in practice.

Robustness

In summary, likelihood methods are based on the assumed model and can perform poorly even when the true model is in a neighborhood of the assumed model. Robustness, on the other hand, aims to protect the statistical methods when the true model is in a neighborhood of the assumed model. The normal model is typically assumed in practice, and there is virtually no way, with moderate sample sizes, to determine from the data that the true model is different but in a neighborhood of the normal model. Researchers often go ahead and use the non-robust normal likelihood methods.

Huber's resolution is to truncate the normal likelihood equation and produce $\hat{\theta}$ and $\Sigma \psi(y_i)$ as estimate and test statistic. They behave quite well over the neighborhood at a slight loss in efficiency at the assumed model, a minimal price for the resulting robustness. The estimator $\hat{\theta}$ has bounded influence and .5 breakdown point, both of which are more favorable than the normal likelihood estimate $\bar{y}$.

Hence, Huber's ψ function becomes our starting point in developing robust methods in the important regression model. Before moving on to the regression model, we briefly mention other approaches to constructing robust procedures. Linear combinations of order statistics have received much attention. These include the trimmed mean, along with the median as a special case. These estimates have bounded influence and positive breakdown points; see Huber (1981), Hampel et al. (1986), and Staudte and Sheather (1990). Another class of robust estimates can be derived from nonparametric rank tests. This approach is developed in detail in Hettmansperger (1984).

6. THE SIMPLE REGRESSION MODEL

The simple regression model allows us to illustrate most of the important new ideas in robustness. Practical applications commonly involve the multiple regression model. We comment on the extension to multiple regression at the end of this chapter.

We now suppose that $y_i = \beta_0 + x_i\beta_1 + \epsilon_i$, $i = 1, \ldots, n$, where $x_1, \ldots, x_n$ are given values of the explanatory variable, β_0 and β_1 are unknown regression parameters, and $\epsilon_1, \ldots, \epsilon_n$ are i.i.d. errors with density symmetric about 0. In matrix notation we write

$$\mathbf{Y} = X\beta + \epsilon$$

where X is the $n \times 2$ known design matrix and β is the 2×1 vector $(\beta_0, \beta_1)^T$. The matrix notation generalizes immediately to multiple regression.

Estimation

The estimating equations for M-estimates are now:

$$\Sigma\psi\left(\frac{y_i - \beta_0 - x_i\beta_1}{\sigma}\right) = 0$$

$$\Sigma\psi\left(\frac{y_i - \beta_0 - x_i\beta_1}{\sigma}\right)x_i = 0$$

Let $w_i(\beta_0, \beta_1) = \min\{1, k\sigma/|y_i - \beta_0 - x_i\beta_1|\}$. Then the equations become:

$$\Sigma w_i(\beta_0, \beta_1)\left(\frac{y_i - \beta_0 - x_i\beta_1}{\sigma}\right) = 0$$

$$\Sigma w_i(\beta_0, \beta_1)\left(\frac{y_i - \beta_0 - x_i\beta_1}{\sigma}\right)x_i = 0.$$

Taking $w_i = 1$ yields the normal likelihood equations, which are also the least-squares normal equations.

Computing the Estimates

We now discuss the iterative computation of the solution $\hat{\beta}_0, \hat{\beta}_1$ and, at the same time, discuss how to handle the estimation of σ.

1. First we must determine starting values $\hat{\beta}_0^{(0)}$ and $\hat{\beta}_1^{(0)}$ and $\hat{\sigma}^{(0)}$. This may be done as follows: First compute the regular least-squares estimates $\hat{\beta}_1^{(0)} = \Sigma(x_i - \bar{x})y_i/\Sigma(x_i - \bar{x})^2$ and $\hat{\beta}_0^{(0)} = \bar{y} - \bar{x}\hat{\beta}_1^{(0)}$. Then form the least-squares residuals $r_i^0 = y_i - \hat{\beta}_0^{(0)} - x_i\hat{\beta}_1^{(0)}$ and find $\hat{\sigma}^{(0)} = 1.483$ MAD $= 1.483$ med$|r_i^0 - $ med $r_j^0|$.

2. Abbreviate $w_i(\beta_0, \beta_1)$ to w_i and let $W_i = w_i/\Sigma w_j$. Further, let $\bar{y}_w = \Sigma W_i y_i$ and $\bar{x}_w = \Sigma W_i x_i$. Then compute $w_i^{(0)} = \min\{1, k\hat{\sigma}^{(0)}/|r_i^0|\}$, and the solutions to the weighted least-squares equations are

$$\hat{\beta}_1^{(1)} = \Sigma w_i^{(0)}(x_i - \bar{x}_w)y_i/\Sigma w_i^{(0)}(x_i - \bar{x}_w)^2$$

$$\hat{\beta}_0^{(1)} = \bar{y}_w - \bar{x}_w\hat{\beta}_1^{(1)}.$$

3. Now using $\hat{\beta}_0^{(1)}, \hat{\beta}_1^{(1)}, r_i^{(1)} = y_i - \hat{\beta}_0^{(1)} - x_i\hat{\beta}_1^{(1)}$ and retaining $\hat{\sigma}^{(0)}$, compute $\hat{\beta}_0^{(2)}$ and $\hat{\beta}_1^{(2)}$ as in step 2.

4. Continue until $\left|\hat{\beta}_0^{(j+1)} - \hat{\beta}_0^{(j)}\right|$ and $\left|\hat{\beta}_1^{(j+1)} - \hat{\beta}_1^{(j)}\right|$ are both small; then take $\hat{\beta}_0 = \hat{\beta}_0^{(j)}$, $\hat{\beta}_1 = \hat{\beta}_1^{(j)}$.

5. Now update $\hat{\sigma}$ by letting $r_i = y_i - \hat{\beta}_0 - x_i\hat{\beta}_1$ and then update $\hat{\beta}_0$ and $\hat{\beta}_1$ once more using step 2. The final estimate of scale is $\hat{\sigma} = 1.483$ MAD, computed from $r_1, \ldots, r_n$, the final residuals.

In order to ensure convergence, Huber (1981, p. 176) has recommended not to iterate the calculation of $\hat{\sigma}$. Many statistical packages such as Minitab and SAS have the capacity to carry out least-squares and weighted least-squares calculations. In step 1 we compute and store the least-squares residuals and compute $\hat{\sigma}^{(0)}$. In step 2 we compute the weights $w_i^{(0)}$ and use these in a weighted least-squares program to find $\hat{\beta}_0^{(1)}$ and $\hat{\beta}_1^{(1)}$ and $r_1^{(1)}, \ldots, r_n^{(1)}$. It is not necessary to compute $w_i^{(0)}$; the program does this automatically. The appendix provides Minitab macros for these calculations for the Huber ψ function.

Asymptotic Distribution

A linear approximation, similar to Proposition 1, is the key to determining the asymptotic distribution of the estimates and also to developing an approximate size α test of $H_0 : \beta_1 = 0$ vs. $H_A : \beta_1 \neq 0$. We suppose for convenience that the true parameter values are $\beta_0 = 0$ and $\beta_1 = 0$ and make the expansions around the true values. We then have

$$\sum_{i=1}^{n} \psi \left(\frac{y_i - \beta_0 - x_i\beta_1}{\sigma} \right) \doteq \sum_{i=1}^{n} \psi \left(\frac{y_i}{\sigma} \right) .$$
$$-\beta_0 \sum_{i=1}^{n} \frac{1}{\sigma} \psi' \left(\frac{y_i}{\sigma} \right) - \beta_1 \sum_{i=1}^{n} \frac{1}{\sigma} \psi' \left(\frac{y_i}{\sigma} \right) x_i$$

$$\sum_{i=1}^{n} \psi \left(\frac{y_i - \beta_0 - x_i\beta_1}{\sigma} \right) x_i \doteq \sum_{i=1}^{n} \psi \left(\frac{y_i}{\sigma} \right) x_i$$
$$-\beta_0 \sum_{i=1}^{n} \frac{1}{\sigma} \psi' \left(\frac{y_i}{\sigma} \right) x_i - \beta_1 \sum_{i=1}^{n} \frac{1}{\sigma} \psi' \left(\frac{y_i}{\sigma} \right) x_i^2 .$$

From these equations, the approximation reduces, as in Proposition 1, to

$$n^{-1/2} \sum_{i=1}^{n} \psi \left(\frac{y_i - \beta_0 - x_i\beta_1}{\sigma} \right) \doteq n^{-1/2} \sum_{i=1}^{n} \psi \left(\frac{y_i}{\sigma} \right)$$
$$-n^{1/2}\beta_0 \frac{1}{\sigma} E\psi' - n^{1/2}\beta_1 \frac{1}{\sigma} E\psi' \bar{x}$$

$$n^{-1/2} \sum_{i=1}^{n} \psi \left(\frac{y_i - \beta_0 - x_i\beta_1}{\sigma} \right) x_i \doteq n^{-1/2} \sum_{i=1}^{n} \psi \left(\frac{y_i}{\sigma} \right) x_i$$
$$-n^{1/2}\beta_0 \frac{1}{\sigma} E\psi' \bar{x} - n^{1/2}\beta_1 \frac{1}{\sigma} E\psi' \frac{1}{n} \sum_{i=1}^{n} x_i^2 .$$

The notation $E\psi'$ means $\int \psi'(t/\sigma) f(t) dt$, where f is the density of the underlying error distribution. Similarly, we use $E\psi^2$ to mean $\int \psi^2(t/\sigma) f(t) dt$. We could make the underlying scale explicit by writing $f(t) = \sigma^{-1} f_1(\sigma^{-1}t)$, where f_1 has scale $\sigma = 1$.

Without loss of generality we assume that the x variable has mean 0, or has been centered. This can be accomplished from the original model as follows: $y_i = \beta_0 + x_i\beta_1 + \epsilon_i = (\beta_0 + \bar{x}\beta_1) + (x_i - \bar{x})\beta_1 + \epsilon_i$. Hence, the intercept becomes $\beta_0^* = \beta_0 + \bar{x}\beta_1$, but the slope parameter is unchanged.

With the assumption that $\bar{x} = 0$ (centering carried out if necessary), we set the equations equal to 0 and solve the linear approximations to find

$$n^{1/2} \hat{\beta}_0 \doteq \frac{\sigma}{E\psi'} n^{-1/2} \sum_{i=1}^{n} \psi \left(\frac{y_i}{\sigma} \right)$$

$$n^{1/2} \hat{\beta}_1 \doteq \frac{\sigma}{E\psi'} \frac{1}{n^{-1}\Sigma x_i^2} n^{-1/2} \sum_{i=1}^{n} \psi \left(\frac{y_i}{\sigma} \right) x_i .$$

By this we mean that the estimates $n^{1/2} \hat{\beta}_0$ and $n^{1/2} \hat{\beta}_1$ can be approximated by solutions to the linear approximation up to $o_p(1)$, as in Propositions 1 and 2. This is convenient because we now have sums of independent random variables, and these can be handled by the central limit theorem. Further, recall that the true parameter values are $\beta_0 = 0$ and $\beta_1 = 0$.

Conditions must be imposed on the design in order to establish the asymptotic normality of the estimators. For example, if we assume that $n^{-1}\Sigma x_i^2 \to \tau^2 > 0$ and $n^{-1} \max x_i^2 \to 0$, then the Lindeberg central limit theorem (Staudte and Sheather 1990, p. 290) yields for the estimate of slope

$$n^{1/2} \hat{\beta}_1 \xrightarrow{D} N \left(0, \frac{\sigma^2 E\psi^2}{(E\psi')^2} \tau^{-2} \right) ;$$

i.e., $\hat{\beta}_1$ is approximately normal with mean 0 and variance $[(E\psi^2)/(E\psi')^2]\sigma^2/\Sigma x_i^2$. In addition, $\hat{\beta}_0$ is approximately normal with mean 0 and variance $\sigma^2 E\psi^2/(E\psi')^2$, and, moreover, $n^{1/2}\hat{\beta}_0$ and $n^{1/2}\hat{\beta}_1$ are jointly asymptotically bivariate normal and uncorrelated because $\bar{x} = 0$.

Finally, note that, if we are using the normal likelihood estimates (least squares), then under the same conditions the same limiting distributions arise with $E\psi^2/(E\psi')^2$ replaced by 1. That is, the variance of the least-squares estimate of β_1 is $\sigma^2/\Sigma x_i^2$; see Arnold (1990, p. 409), for example. The limiting distribution is also exact when the underlying distribution is normal.

In practice $E\psi^2$, $E\psi'$, and σ must be estimated from the data. Earlier we pointed out that $\hat{\sigma} = 1.483$ MAD, computed from $r_i = y_i - \hat{\beta}_0 - x_i\hat{\beta}_1$. In addition, $E\psi^2$ is estimated by $(n-2)^{-1} \sum_{i=1}^{n} \psi^2(r_i/\hat{\sigma})$ and $E\psi'$ is estimated by $n^{-1} \sum_{i=1}^{n} \psi'(r_i/\hat{\sigma})$. The factor $(n-2)^{-1}$ was suggested by Huber (1981, p. 173) to help control the bias.

Testing

We now turn to a test of $H_0 : \beta_1 = 0$ vs. $H_A : \beta_1 \neq 0$. The test statistic is assembled in two steps:

1. Assuming the null hypothesis $H_0 : \beta_1 = 0$ is true, we estimate β_0 with β_0^*, the solution to $\Sigma\psi[(y_i - \beta_0)/\hat{\sigma}] = 0$, where $\hat{\sigma}$ is the estimate of σ based on the full-model residuals.

2. Then the test statistic is $S = \Sigma\psi \left(\frac{y_i - \beta_0^*}{\hat{\sigma}} \right) x_i$, and we reject $H_0 : \beta_1 = 0$ if $|S| > z_{\alpha/2} (\hat{E}\psi^2 \Sigma x_i^2)$, where

$\hat{E}\psi^2$ is the estimate of $E\psi^2$ and $1 - \Phi(z_{\alpha/2}) = \alpha/2$. This test has approximately size α because

$$n^{-1/2} \sum_{i=1}^{n} \psi\left(\frac{y_i - \beta_0^*}{\hat{\sigma}}\right) x_i \doteq$$

$$n^{-1/2} \sum_{i=1}^{n} \psi\left(\frac{y_i - \beta_0}{\sigma}\right) x_i$$

from the linear approximation and $\bar{x} = 0$. Then $n^{-1/2} \Sigma \psi[(y_i - \beta_0)/\sigma] x_i$ converges in distribution under $H_0 : \beta_1 = 0$ to normality with mean 0 and variance $E\psi^2 \tau^2$, where $n^{-1} \Sigma x_i^2 \to \tau^2 > 0$.

A second possibility, more direct, is to reject $H_0 : \beta_1 = 0$ if $|\hat{\beta}_1| > z_{\alpha/2} \hat{\sigma}[(\hat{E}\psi^2)^{1/2}/\hat{E}\psi']/[\Sigma x_i^2]^{1/2}$. This test also has approximately size α because of the limiting normality of $n^{1/2} \hat{\beta}_1$ discussed above.

Example 4. We consider the data given by Moore and McCabe (1989, p. 183). These data are part of a study of the cognitive development of children. The dependent variable y is the score on a verbal aptitude test given to a child long after he has begun to speak, and the independent variable x is the age in months at which the child spoke his first word. The data are listed in Table 5 in Section 8. One issue of interest is whether aptitude can be predicted from age at first speech. A scatter plot of the data, along with the least-squares regression line, is discussed in Chapter 1 of this volume. Observations 18 and 19 are unusual and stand out in the plot. They exert influence on the least-squares line.

We use these data to illustrate how normal likelihood methods (least squares) are influenced by observations 18 and 19. The Huber ψ function will help in the case of observation 19. We will need a more elaborate method to deal with observation 18, and we discuss this method in the next section and illustrate it in Example 5. Using Minitab, we find the normal likelihood solution is

$$\hat{y} = 109.874 - 1.1270x.$$

In order to compute the Huber estimates we use the Minitab macros given in the appendix. Beginning with the normal likelihood estimates, the macros first compute the Huber one-step estimates and then iterate. The user sets the number of iterations and monitors the maximum absolute difference between the last two sets of Huber estimates, stopping when this difference is sufficiently small. The macros are interactive, and the user inputs the number of variables plus one (the dimension

of the problem), 2 in this example, and the value of k, 1.28 in this example.

The output includes, along with the estimates and differences from the previous iteration, the weights in the weighted least-squares equations being solved. These weights are informative because they show which observations are downweighted, and the researcher should inspect these observations for anomalies. At the final iteration, the estimated variance-covariance matrix and standard errors are printed. For this data set Table 4 shows the iterations.

Hence

$$\hat{y} = 110.488 - 1.202x$$

is the prediction equation, different from the earlier least-squares prediction equation. After four iterations the maximum absolute difference from the previous estimates is 0.0016. At this point $\hat{\sigma}$ is updated, and row 5 in the table shows the final Huber estimates after one more iteration with the new value of $\hat{\sigma}$. The final weights are 1.000, 1.000, 0.724, 1.000, 1.000, 1.000, 1.000, 1.000, 1.000, 1.000, 0.971, 1.000, 0.724, 0.845, 1.000, 1.000, 1.000, 1.000, 0.353, 0.999, 1.000. Note that observation number 19 is strongly downweighted, whereas number 18 is not downweighted at all. This happens because number 19 does not correspond to an extreme value of x, whereas number 18 is an extreme observation. Observations with extreme x values will be discussed in the next section. Observations 3, 13, and 14 should be inspected also because they too have been downweighted.

The estimated variance-covariance matrix is

$$\begin{bmatrix} 17.4638 & -0.6465 \\ -0.6465 & 0.0318 \end{bmatrix}$$

and the estimated standard errors of $\hat{\beta}_0$ and $\hat{\beta}_1$ are 4.180 and 0.178, respectively. The standard errors are the square roots of the diagonal elements of the variance-covariance matrix.

The test of $H_0 : \beta_1 = 0$ can be based on $t^* = \hat{\beta}_1/\text{s.e.}(\hat{\beta}_1) = -1.202/0.178 = -6.753$, which is more extreme than the least-squares $t = -3.63$.

We would use the standard normal distribution to approximate the P-value of both t and t^*. The P-value is essentially 0 in both instances. However, these excessively small P-values may be caused by observation 18. A very different view is presented in Example 5.

Iteration	$\hat{\beta}_0$	$\hat{\beta}_1$	
Start	109.874	−1.127	Normal Likelihood Estimates
1	110.052	−1.181	Huber One-Step Estimates
2	109.959	−1.178	
3	109.946	−1.178	
4	109.945	−1.178	
5	110.488	−1.202	Final Huber Estimates

Table 4: Iterations for Huber estimates in Example 4.

Influence

We now consider the influence function for $\hat{\beta}_1$, the estimate of slope. Following the same heuristic development as for the location estimate discussed previously, we suppose that $\hat{\beta}_{1,n} = 0$ and then we add an $(n+1)$th observation (y^*, x^*). We wish to estimate the differential change in $\hat{\beta}_1$. We see at once from the linear approximation that

$$(n+1)\,\hat{\beta}_{1,n+1} \doteq \frac{\sigma}{\tau^2 E\psi'}\,\psi\left(\frac{y^*}{\sigma}\right) x^* = \Omega\,(y^*, x^*).$$

Clearly, the influence function $\Omega\,(y^*, x^*)$ is bounded in y^* but unbounded in x^*. Hence, we have lost the qualitative robustness that the same ψ function gave for the location estimate. The breakdown point is also 0. This is not a problem in a well-designed experiment, in which the x values are carefully chosen. We then have protection against outliers in the dependent variable y. However, if we are conditioning the analysis on observed x values, then outlying x values present a distinct problem.

To cope with this threat, in the next section we use ideas from regression diagnostics. For the normal likelihood methods (least squares), these diagnostic techniques aid in detecting anomalous data, both in y and in x values. Using the results developed for diagnostics, we describe how to modify the method based on ψ to produce a bounded-influence estimate of the regression parameter β_1.

7. DIAGNOSTICS

We begin this section with a review of some basic material in regression diagnostics. For additional reading see Chapter 8 in this volume and its references.

Leverage

Because we have taken $\bar{x} = 0$, the normal likelihood estimates of β_0, β_1 are $\hat{\beta}_{0,LS} = \bar{y}, \hat{\beta}_{1,LS} = \Sigma x_j y_j / \Sigma x_k^2$. Hence the ith predicted (fitted) value is

$$
\begin{aligned}
\hat{y}_i &= \hat{\beta}_{0,LS} + x_i \hat{\beta}_{1,LS} \\
&= \bar{y} + x_i \Sigma x_j y_j / \Sigma x_k^2 \\
&= \sum_{j=1}^{n} \left(\frac{1}{n} + \frac{x_i x_j}{\Sigma x_k^2}\right) y_j \\
&= \sum_{j=1}^{n} h_{ij} y_j
\end{aligned}
$$

where $h_{ij} = n^{-1} + x_i x_j / \Sigma x_k^2$. The quantity $h_{ii} = n^{-1} + x_i^2 / \Sigma x_k^2$ is called the ith *leverage value* because it determines how much impact y_i has in determining its own predicted value, $\hat{y}_i$. As h_{ii} approaches 1, $\hat{y}_i \doteq y_i$ and the regression line is levered toward the ith point. Leverage is a function of the dispersion of the x-values, which are centered at 0. A large x_i (relative to Σx_k^2) implies that h_{ii} is large. Hence, points that are outlying in x exert a strong influence on the regression line. This diagnostic is especially important in multiple regression involving more than two x-variables because then the x-variables cannot be plotted all together.

Residuals

The ith residual is

$$r_i = y_i - \hat{y}_i = (1 - h_{ii})\,y_i - \sum_{j \neq i} h_{ij}\,y_j.$$

Here again, when h_{ii} is large, $1 - h_{ii}$ is small and the h_{ij} are all small, so that $r_i \doteq 0$.

It can be shown that var $\hat{y}_i = \sigma^2 h_{ii}$ and var $r_i = \sigma^2(1 - h_{ii})$. We usually assess the size of a residual relative to its standard deviation. The *standardized least-squares residual* is $r_i^* = r_i / \sigma\,(1 - h_{ii})^{1/2}$. Provided we have some estimate $\hat{\sigma}$, one of the most informative graphical displays is a plot of the standardized residuals versus the predicted values. If the regression model is adequate, this plot should be a random scatter with most values of r_i^* within 2 or so units of 0. Patterns in this plot suggest such departures as error variance that increases or decreases with x or a need to transform the data to achieve a good fit. We are interested primarily in observations for which $|r_i^*| > 2$. These suggest points where the regression equation does not fit well. Cook and Weisberg (1982, pp. 18–19) and Belsley, Kuh, and Welsch (1980, pp. 19–20) provide extensive discussions of this diagnostic with different estimates of σ.

A Combined Measure

The ratio $h_{ii}/(1 - h_{ii})$, called *potential* by Cook and Weisberg (1982, p. 117), varies from 0 to ∞ as h_{ii} varies from 0 to 1. Combining potential with r_i^* produces the diagnostic

$$d_i = r_i^* \left(\frac{h_{ii}}{1 - h_{ii}} \right)^{1/2}.$$

The diagnostic d_i responds to either high leverage (extreme x value) or a large residual (point far from fitted line) and thus combines both important aspects into a single quantity. This quantity has been studied extensively with different estimates of σ in r_i^*. See Staudte and Sheather (1990, Section 7.2), Cook and Weisberg (1982, Section 3.5.1), and Belsley et al. (1980, Section 2.1).

Diagnostics in Robust Regression

The measures r_i^* and d_i have been developed in the context of normal likelihood (least-squares) methodology. Our purpose here is to adapt them for use with robust methods and to use them to further modify ψ to achieve bounded influence. We have two obvious options: first to robustify r_i^*, and perhaps d_i, by replacing LS estimates by M-estimates, or second, to go back to basics and re-define at least r_i^* to use the proper standard deviation of the M-estimate in the denominator. It is a convenient feature of normal likelihood methods that the leverage value h_{ii} appears in the calculation of $\hat{y}_i$ and r_i and also in the variance expressions. We cannot expect this to happen for the fitted values and residuals from a robust methodology. We explore this issue first.

Recall the pieces that we have assembled so far by using ψ. We have M-estimators $\hat{\beta}_0$ and $\hat{\beta}_1$, which then yield $\hat{y}_i = \hat{\beta}_0 + x_i\hat{\beta}_1$ and $r_i = y_i - \hat{\beta}_0 - x_i\hat{\beta}_1$. Recall further

that, from the linear approximation in Section 6,

$$\hat{\beta}_0 \doteq \frac{\sigma}{E\psi'} \frac{1}{n} \Sigma\psi\left(\frac{y_i}{\sigma}\right)$$

$$\hat{\beta}_1 \doteq \frac{\sigma}{E\psi'} \frac{\Sigma\psi(y_i/\sigma)x_i}{\Sigma x_i^2}.$$

Using these approximations, we can develop first-order approximations to the variances of the predicted values and residuals based on M-estimates.

Proposition 6:

$$\text{var } \hat{y}_i \doteq \frac{\sigma^2 E\psi^2}{(E\psi')^2} h_{ii}$$

$$\text{var } r_i \doteq \sigma^2(1 - K h_{ii})$$

where

$$K = \frac{2E[(\epsilon/\sigma)\,\psi(\epsilon/\sigma)]}{E\psi'} - \frac{E\psi^2}{(E\psi')^2},$$

and ϵ represents the error random variable.

We will sketch the derivation of var r_i. Replace $\hat{\beta}_0$ and $\hat{\beta}_1$ by their approximations to get:

$$r_i \doteq y_i - \frac{\sigma}{E\psi'} \frac{1}{n} \Sigma\psi\left(\frac{y_j}{\sigma}\right) - x_i \frac{\sigma}{E\psi'} \frac{\Sigma\psi(y_j/\sigma)x_j}{\Sigma x_j^2}$$

$$= y_i - \sum \left\{ \frac{\sigma}{nE\psi'} + x_ix_j \frac{\sigma}{E\psi'\Sigma x_j^2} \right\} \psi(y_j/\sigma)$$

$$= y_i - \frac{\sigma}{E\psi'} \left(\frac{1}{n} + \frac{x_i^2}{\Sigma x_j^2} \right) \psi(y_i/\sigma)$$

$$\quad - \frac{\sigma}{E\psi'} \sum_{j \neq i} \left(\frac{1}{n} + \frac{x_jx_i}{\Sigma x_j^2} \right) \psi(y_j/\sigma)$$

$$= \left[y_i - \frac{\sigma}{E\psi'} h_{ii}\,\psi(y_i/\sigma) \right]$$

$$\quad - \frac{\sigma}{E\psi'} \sum_{j \neq i} h_{ij}\psi(y_j/\sigma)$$

$$\text{var } r_i \doteq E\left[Y_i - \frac{\sigma}{E\psi'} h_{ii}\,\psi(Y_i/\sigma) \right]^2$$

$$\quad + \frac{\sigma^2}{(E\psi')^2} \sum_{j \neq i} h_{ij}^2\, E\psi^2(Y_j/\sigma).$$

We have assumed that the true values are $\beta_0 = 0$ and $\beta_1 = 0$ without loss of generality, that $Y_1, \ldots, Y_n$ are independent with mean 0 and variance σ^2, and also that

$E\psi(Y_i/\sigma) = 0$. So we have

$$\text{var } r_i \;\doteq\; \sigma^2 - \frac{2\sigma}{E\psi'}\, h_{ii} E[Y_i\psi(Y_i/\sigma)]$$

$$+ \frac{\sigma^2}{(E\psi')^2}\, h_{ii}^2\, E\psi^2 + \frac{\sigma^2}{(E\psi')^2}\, E\psi^2 \sum_{j\neq i} h_{ij}^2.$$

Finally use $h_{ii} = \sum_j h_{ij}^2$ and the result follows.

Hence, we can now standardize a residual from a robust regression based on ψ as

$$r_i^* = \frac{y_i - \hat\beta_0 - x_i\hat\beta_1}{\hat\sigma\,(1 - \hat{K}h_{ii})^{1/2}}$$

with σ, $E\psi^2$ and $E\psi'$ estimated as described above and $E[(\epsilon/\sigma)\,\psi\,(\epsilon/\sigma)]$ estimated by $n^{-1}\Sigma(r_j/\hat\sigma)\,\psi(r_j/\hat\sigma)$. The estimate $\hat{K}$ of K can then be assembled from these pieces. We will continue to use r_i^* to represent a standardized residual. If the estimates are least-squares estimates, then we have the earlier diagnostic.

The leverage $h_{ii} = n^{-1} + x_i^2/\Sigma x_j^2$ could also be replaced by a robust measure of distance in x space. This is strongly advocated by Rousseeuw and Leroy (1987, p. 265). However, because alternatives are not yet readily computable or available, we retain h_{ii} as defined. Hence, we continue to use

$$d_i = r_i^* \left(\frac{h_{ii}}{1 - h_{ii}}\right)^{1/2}.$$

Now $\Sigma h_{jj} = 2$, so the average leverage is $2/n$. Values of $|r_i^*|$ greater than 2 generally attract attention; hence, we might consider d_i excessive if $|d_i| > 2[(2/n)/(1 - 2/n)]^{1/2} = 2[2/(n-2)]^{1/2}$. A simpler rule that seems to work in practice is $|d_i| > 2.0\,(2/n)^{1/2}$. A plot of d_i versus observation number or versus $\hat{y}_i$ helps find unusual observations.

In many practical situations, we have found K to be close to 1. For example, if f is a standard normal density, then $EY\psi(Y) = \int t\psi(t)\,f(t)\,dt$. But $tf(t) = -f'(t)$, so we have $-\int\psi(t)\,f'(t)\,dt = \int\psi'(t)\,f(t)\,dt = E\psi'$ with an integration by parts. The first term in K is equal to 2, and $K = 2 - E\psi^2/(E\psi')^2$. But k is chosen so that $E\psi^2/(E\psi')^2$ is greater than .9. Hence K is close to 1, and in constructing r_i^*, we could ignore $\hat{K}$. On the other hand, it is important to use robust estimates for β_0, β_1 and σ.

In summary, we have developed diagnostics similar to those used with normal likelihood (least-squares) meth-

ods. They provide invaluable plots to help detect problems with the model or anomalies in the data, as Example 5 illustrates. Our next task is to circle back on our old arguments and try to use diagnostic quantities to construct bounded-influence methods.

8. BOUNDED-INFLUENCE REGRESSION ESTIMATES

Recall from Section 6 that the M-estimates $\hat\beta_0, \hat\beta_1$ are actually found by using the method of weighted least squares. The weights are, with $r_i = y_i - \hat\beta_0 - x_i\hat\beta_1$,

$$w_i(\hat\beta_0, \hat\beta_1) = \min\left\{1, \frac{k\hat\sigma}{|r_i|}\right\}.$$

From the point of view of our diagnostics, either least-squares or robust, we might do better to use

$$\min\left\{1, \frac{k}{|r_i^*|}\right\}$$

but this modification corrects only the standardization of the residual. This approach would downweight observations smoothly in proportion to the properly standardized residual. Let us go further and consider

$$\min\left\{1, \frac{k}{|d_i|}\right\} = \min\left\{1, \frac{k}{|r_i^*|}\left(\frac{1 - h_{ii}}{h_{ii}}\right)^{1/2}\right\}.$$

This choice will smoothly downweight observations in proportion to the excess of $|d_i|$ over k. This approach allows downweighting of high leverage observations that do not fit well and will not downweight an observation if it does fit well. It combines the size of standardized residual and the size of the leverage. We will concentrate on using the least-squares diagnostics, but, the robust versions could be used also.

Using the Modified Weights

The question now is what happens to our original equations based on Huber's ψ? Using the least-squares r_i^*, we write

$$w_i(\hat\beta_0, \hat\beta_1) = \min\left\{1, \frac{k\hat\sigma v_i}{|r_i|}\right\}$$

where $v_i = (1 - h_{ii})/h_{ii}^{1/2}$. We refer to these weights as *Welsch weights*; Welsch (1980) proposed a special case of these weights. We discuss this weight in more detail in Example 5. It is easy to check that the appropriate equations in terms of ψ are

$$\Sigma \sigma v_i \psi \left(\frac{y_i - \beta_0 - x_i \beta_1}{\sigma v_i} \right) = 0$$

$$\Sigma \sigma v_i \psi \left(\frac{y_i - \beta_0 - x_i \beta_1}{\sigma v_i} \right) x_i = 0.$$

We discuss the computation of the solutions $\hat{\beta}_0$ and $\hat{\beta}_1$ in Example 5.

We need to check whether the influence function of $\hat{\beta}_1$ is bounded. Again, suppose the true values are $\beta_0 = 0$ and $\beta_1 = 0$. Then linear approximations to the above equations yield:

$$\Sigma \sigma v_i \psi \left(\frac{y_i}{\sigma v_i} \right) - \Sigma \psi' \left(\frac{y_i}{\sigma v_i} \right) \beta_0$$
$$- \Sigma \psi' \left(\frac{y_i}{\sigma v_i} \right) x_i \beta_1 = 0$$
$$\Sigma \sigma v_i \psi \left(\frac{y_i}{\sigma v_i} \right) x_i - \Sigma \psi' \left(\frac{y_i}{\sigma v_i} \right) x_i \beta_0$$
$$- \Sigma \psi' \left(\frac{y_i}{\sigma v_i} \right) x_i^2 \beta_1 = 0$$

The factor v_i in $y_i/(\sigma v_i)$ locks the ψ' into the sum, and we cannot simplify these expressions with $\bar{x} = 0$ as we did in Section 6.

Approximate Distribution

These linear approximations can be used for two results: first to construct the approximating distributions of $\hat{\beta}_0$ and $\hat{\beta}_1$ and then to find the influence function. It is easier to continue the development in matrix notation.

Let $\mathbf{x}_i^T = (1 \; x_i)$ and $\beta^T = (\beta_0 \; \beta_1)$. Then the equations can be written compactly as

$$\Sigma \sigma v_i \psi \left(\frac{y_i - \mathbf{x}_i^T \beta}{\sigma v_i} \right) \mathbf{x}_i = \mathbf{0}$$

and the linear approximation yields

$$\Sigma \sigma v_i \psi \left(\frac{y_i}{\sigma v_i} \right) \mathbf{x}_i - \Sigma \psi' \left(\frac{y_i}{\sigma v_i} \right) \mathbf{x}_i \mathbf{x}_i^T \beta = \mathbf{0}.$$

Now suppose that $n^{-1} \Sigma \psi'(Y_i/\sigma v_i) \mathbf{x}_i \mathbf{x}_i^T$ converges in probability to a positive-definite matrix M. Then

$$n^{1/2} \hat{\beta} \doteq M^{-1} n^{-1/2} \Sigma \sigma v_i \psi \left(\frac{y_i}{\sigma v_i} \right) \mathbf{x}_i.$$

Now, under regularity conditions given by Maronna and Yohai (1981), it can be shown that

$$n^{-1/2} \sigma \Sigma v_i \psi \left(\frac{y_i}{\sigma v_i} \right) \mathbf{x}_i \xrightarrow{D} MVN \left(\mathbf{0}, \sigma^2 Q \right)$$

where $n^{-1} \Sigma v_i^2 \psi^2(Y_i/\sigma v_i) \mathbf{x}_i \mathbf{x}_i^T$ converges in probability to the positive-definite matrix Q. The strategy is the same as for the earlier Huber estimate, but now we must take care to ensure that a central limit theorem will apply under appropriate conditions. Note that the linear approximation can be used to anticipate the correct asymptotic distribution.

Hence, we can check that

$$n^{1/2} \hat{\beta} \xrightarrow{D} MVN \left(\mathbf{0}, V \right) \quad \text{with} \quad V = \sigma^2 M^{-1} Q M^{-1}.$$

If $v_i \equiv 1$, $\sigma^2 M^{-1} Q M^{-1} = \sigma^2 E\psi^2/(E\psi')^2 \Lambda^{-1}$, where Λ is the limit of $n^{-1} X^T X$ and X has ith row $\mathbf{x}_i^T = (1 \; x_i)$. So the results in Section 6 are a special case of the present discussion.

Influence Function

Recall that the influence function at the added point $(y^*, \mathbf{x}^*)$ is determined by the linear approximation. Similar to the earlier case at the end of Section 6, the vector influence is given by

$$\boldsymbol{\Omega}(y^*, \mathbf{x}^*) \doteq \sigma \psi \left(\frac{y^* - \beta_0 - \beta_1 x^*}{\sigma} \right) v^* M^{-1} \mathbf{x}^*.$$

Now $\mathbf{x}^* = (1 \; x^*)^T$ and $v^* = (1 - h^*)/(h^*)^{1/2}$, where h^* is the leverage of the new case $(y^*, \mathbf{x}^*)$. When we check to see whether $\|\boldsymbol{\Omega}(y^*, \mathbf{x}^*)\| = [\boldsymbol{\Omega}^T(y^*, \mathbf{x}^*)\boldsymbol{\Omega}(y^*, \mathbf{x}^*)]^{1/2}$ is bounded, we must look carefully at

$$\begin{aligned}
\boldsymbol{\Omega}^T \boldsymbol{\Omega} &\leq \text{constant} \times (v^*)^2 \times \mathbf{x}^{*T} A \mathbf{x}^* \\
&= \text{constant} \times (v^*)^2 \\
&\quad \times [a_{11} + (a_{12} + a_{21}) x^* + a_{22} x^{*2}]
\end{aligned}$$

where $A = M^{-2}$ and the constant contains σ^2 along with the bound on ψ. We now show that $(v^* x^*)^2$ will be arbitrarily small when x^* is large. Hence $\|\boldsymbol{\Omega}(y^*, \mathbf{x}^*)\|$ is bounded. The design is assumed to be centered so that $\bar{x}_n = 0$. When x^* is added, we have $\bar{x}_{n+1} = x^*/(n+1)$. Hence $h^* = (n+1)^{-1} + (x^* - \bar{x}_{n+1})^2/\Sigma(x_j - \bar{x}_{n+1})^2$, and

$$h^* = \frac{1}{n+1} + \frac{[n/(n+1)]^2 x^{*2}}{\sum_1^n x_j^2 + \left(\frac{n}{n+1} \right) x^{*2}}.$$

	Score (y)	Age (x)	h_{ii}	r_i^*	d_i	w_i
1	95	15	0.048	0.184	0.041	1
2	71	26	0.155	−0.942	−0.403	1
3	83	10	0.063	−1.511	−0.391	1
4	91	9	0.071	−0.814	−0.224	1
5	102	15	0.048	0.833	0.187	1
6	87	20	0.073	−0.031	−0.009	1
7	93	18	0.058	0.311	0.077	1
8	100	11	0.057	0.230	0.056	1
9	104	8	0.080	0.290	0.085	1
10	94	20	0.073	0.618	0.173	1
11	113	7	0.091	1.051	0.332	1
12	96	9	0.071	−0.343	−0.094	1
13	83	10	0.063	−1.511	−0.391	1
14	84	11	0.057	−1.280	−0.314	1
15	102	11	0.057	0.413	0.101	1
16	100	10	0.063	0.127	0.033	1
17	105	12	0.052	0.798	0.187	1
18	57	42	0.652	−0.845	−1.156	0.377
19	121	17	0.053	3.607	0.854	0.725
20	86	11	0.057	−1.076	−0.264	1
21	100	10	0.063	0.127	0.033	1

Table 5: Data and diagnostic values for Example 5.

With a little algebra

$$\frac{(1-h^*)^2}{h^*}\,x^{*2} = \frac{\left(\sum_1^n x_j\right)^2 x^{*2}}{\left(\frac{n+1}{n}\sum_1^n x_j^2 + x^{*2}\right)\left(\frac{1}{n}\sum_1^n x_j^2 + x^{*2}\right)}.$$

For fixed n, as x^* increases, the above quantity decreases, and so is bounded. This is the dominant term so that, because ψ is also bounded, we have achieved our goal of a bounded-influence estimate $\hat\beta = (\hat\beta_0\hat\beta_1)^T$.

Example 5. We now return to Example 4 and illustrate the diagnostics and Welsch estimates. We will use Minitab to carry out the computations. Minitab macros to compute the Welsch estimates and their standard errors are given in the appendix of Staudte and Sheather (1990). We must decide on how to compute r_i^* and d_i. We use least-squares diagnostics for plots and then use these diagnostic values to begin the iterative computations. To estimate σ in r_i^*, we use the root-mean-square error with the ith observation deleted. In this case r_i^* is called the t-resid or (externally) studentized residual, and d_i is called DFITS$_i$, which stands for a scaled change in the ith fitted value when the ith observation is deleted. A discussion of d_i interpreted as a scaled change in the

fitted value can be found in any of the references in Section 7 and will not be pursued here.

In Table 5 we show the data along with the leverage values h_{ii}, the studentized residuals r_i^*, the d_i, and the final Welsch weights w_i. Figure 5(a) is a scatter plot of y (adaptive score) versus x (age at first word) along with the least-squares line, the Huber line, the Welsch line, and the least-squares line without observation 18. Observation 18 and, to a lesser extent, observation 2 show up in the leverage plot in Figure 5(b). Note the similarities between the least-squares, Huber, and Welsch lines. The main difference between the Huber weights in Example 4 and the Welsch weights in Table 5, shown in Figure 6, is that the Welsch estimate severely downweights observation 18, whereas the Huber estimate does not downweight it at all. Thus, the Welsch bounded-influence line is not attracted by the high-leverage observation 18. It is, however, attracted by observation 2, and hence it produces a line similar to both the least-squares line and the Huber line.

Finally, Figure 7 shows plots of r_i^* versus observation number and d_i versus observation number. Note that d_i

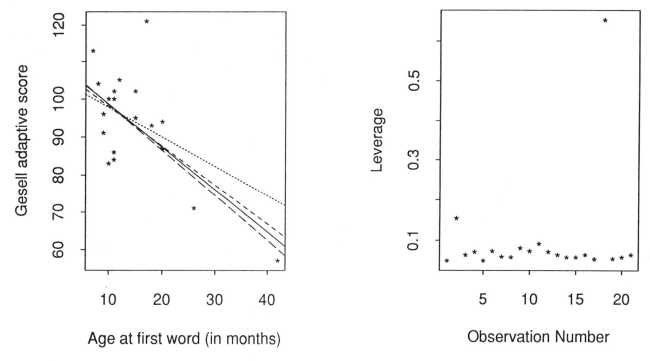

Figure 5: (a) Scatterplot of the Gesell score data with fitted lines. The lines are, from top to bottom at the right of the plot, the least-squares line (without case 18), the Welsch line ($k = 2$), the Huber line ($k = 1.28$), and the least-squares line from all cases. (b) Leverage plot for the Gesell data.

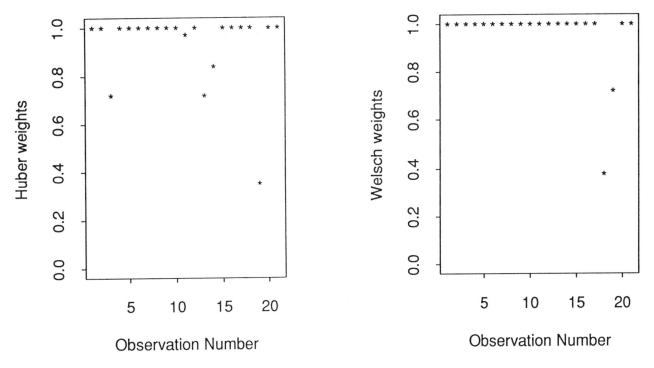

Figure 6: The Huber and Welsch weights for Example 5.

Iteration	$\hat{\beta}_0$	$\hat{\beta}_1$	
Start	109.874	−1.127	Normal Likelihood Estimates
1	108.892	−1.076	Welsch One-Step Estimates
2	108.636	−1.056	
$\vdots$			
8	108.417	−1.037	Final Welsch Estimates

Table 6: Iterations for Welsch estimates in Example 5.

highlights observations 18 and 19 for special attention, whereas r_i^* points only to observation 19. The d_i values are the initial values in the Welsch iterations. The iterations begin with the normal likelihood values and proceed like the Huber iterations described in Example 4. Table 6 shows the results. The value of k for the Welsch weights was taken to be 2.

Hence,

$$\hat{y} = 108.417 - 1.037x$$

is the prediction equation. After 8 iterations, the maximum absolute difference from the estimates in the previous iteration is .00094. The final weights w_i were all 1 except for weights of .377 on observation 18 and .725 on observation 19. The variance-covariance matrix is

$$\begin{bmatrix} 46.1185 & -3.3191 \\ -3.3191 & 0.2714 \end{bmatrix}$$

and the standard errors of $\hat{\beta}_0$ and $\hat{\beta}_1$ are 6.791 and .521, respectively.

The test of $H_0 : \beta_1 = 0$ could be based, as in earlier examples, on $t^* = \hat{\beta}_1/\text{s.e.}(\hat{\beta}_1) = -1.037/.521 = -1.99$. The estimated P-value, using the standard normal distribution, is .047.

Note from Example 4 that both least-squares and Huber estimates have much smaller standard errors and hence yield much smaller P-values than the Welsch estimate. This is because the high-leverage observation number 18 exerts a strong influence and reduces the standard errors of the former estimates. The Welsch weight for this observation is quite small. This means that the Welsch analysis has declared this observation incompatible with the rest of the data and discounted its influence. Recall that the bounded-influence approach is designed to downweight only those observations that are not consistent with the bulk of the data. For this reason, the Welsch analysis is probably more realistic than either the least-squares analysis or the Huber analysis.

Finally, note that if observations 2 and 18 are set aside there is no regression structure left in the data no matter what fitting method we use. To proceed with the analysis of these data we must therefore know if these points are anomalous or if they are in fact good data. This can often be determined by discussion with the scientists who collected the data. If the suspect points are good data, then the robust analysis using Welsch weights is probably the best analysis because the standard errors are more realistic. If they are bad data and must be eliminated, then none of the methods should or would show any regression structure. This example points to the need to use plots to check the results of any analysis, robust or traditional. Hopefully the experimenters collected further data to clarify these issues.

9. EXTENSIONS AND FURTHER DEVELOPMENTS

The multiple regression model with p explanatory variables presents few additional difficulties. The data will appear as $(y_i, 1, x_{i1}, x_{i2}, \ldots, x_{ip})$ for $i = 1, \ldots, n$. Matrix notation is almost essential in this case. The books cited above discuss and illustrate the details of the multiple regression model. The Minitab macros that were used in the example to illustrate simple regression can be used also for multiple regression; see Staudte and Sheather (1990).

The issue of breakdown is more difficult in both the simple and multiple regression models and is beyond the scope of this chapter. In order to develop regression estimators that have 50% breakdown, as the median does in univariate data, Rousseeuw and Leroy (1987) replace the sum of squared residuals by the median of the squared residuals. The resulting estimator has essentially 50% breakdown but unwieldy asymptotic properties. This least-median-of-squares (LMS) estimator has been proposed for use in exploratory data analysis because it is

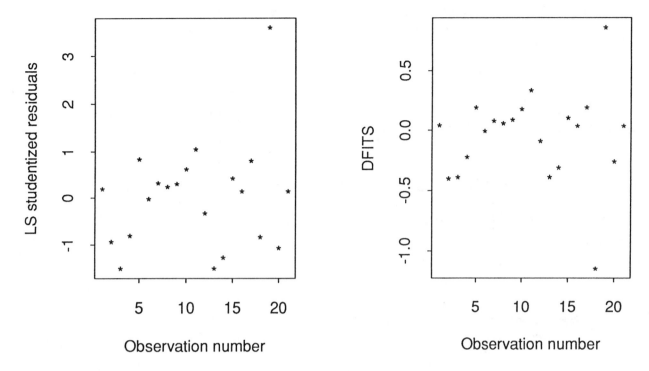

Figure 7: Studentized residuals (left) and DFITS (right) from the least-squares line in Example 5.

so resistant. Rousseeuw and Leroy provide a diskette on request to carry out the calculations. The area of high-breakdown regression estimators is currently under active development in the research literature. The papers in Stahel and Weisberg (1991) provide a comprehensive survey on robustness.

Thomas P. Hettmansperger is Professor of Statistics at the Pennsylvania State University, where he has been on the faculty since receiving his Ph.D. in statistics from the University of Iowa in 1967. His primary research interest is the development and assessment of statistical methods based on ranks. Professor Hettmansperger is a fellow of the American Statistical Association and the Institute of Mathematical Statistics. He has held visiting positions at Berkeley, Princeton, La Trobe University, Melbourne University, University of New South Wales, and University of Bern. In 1986 he was awarded the C. I. Noll College of Science Teaching Award.

Simon J. Sheather is Senior Lecturer at the Australian Graduate School of Management, University of New South Wales. Dr. Sheather has been Lecturer in Statistics at the University of Melbourne and Visiting Associate Professor of Statistics at the Pennsylvania State University. His research interests are in the areas of nonparametric and robust methods. In 1991 Dr. Sheather was the recipient of the Australian Graduate School of Management Alumni Association award for excellence in teaching. He received his Ph.D. in statistics from La Trobe University.

REFERENCES

Arnold, S.F. (1990), *Mathematical Statistics*, Englewood Cliffs, NJ: Prentice-Hall.

Belsley, D.A., Kuh, E., and Welsch, R.E. (1980), *Regression Diagnostics*, New York: Wiley.

Bickel, P.J. and Doksum, K.A. (1977), *Mathematical Statistics: Basic Ideas and Selected Topics*, San Francisco: Holden-Day.

Cook, R.D. and Weisberg, S. (1982), *Residuals and Influence in Regression*, London and New York: Chapman and Hall.

Cushny, A.R. and Peebles, A.R. (1905), "The Action of Optical Isomers. II, Hyoscines," *Journal of Physiology*, **32**, 501–510.

Diaconis, P. and Efron, B. (1983), "Computer-Intensive Methods in Statistics," *Scientific American*, **248**, 5, 116–130.

Efron, B. (1982), *The Jackknife, the Bootstrap, and Other Resampling Plans*, Philadelphia: Society for Industrial and Applied Mathematics.

Fienberg, S.E. and Hinkley, D.V. (eds.) (1980), *R.A. Fisher: An Appreciation*, New York: Springer-Verlag.

Hampel, F.R., Ronchetti, E.M., Rousseeuw, P.J., and Stahel, W.A. (1986), *Robust Statistics: The Approach Based on Influence Functions*, New York: Wiley.

Hettmansperger, T.P. (1984), *Statistical Inference Based on Ranks*, New York: Wiley.

Hoaglin, D.C. and Welsch, R.E. (1978), "The Hat Matrix in Regression and ANOVA," *The American Statistician*, **32**, 17–22, 146 (correction).

Huber, P.J. (1964), "Robust Estimation of a Location Parameter," *Annals of Mathematical Statistics*, **35**, 73–101.

—— (1981), *Robust Statistics*, New York: Wiley.

Maronna, R.A. and Yohai, V.J. (1981), "Asymptotic Behavior of General M-estimates for Regression and Scale with Random Carriers," *Zeitschrift für Wahrscheinlichkeitstheorie und verwandte Gebiete*, **58**, 7–20.

Moore, D.S. and McCabe, G.P. (1989), *Introduction to the Practice of Statistics*, New York: Freeman.

Pearson, E.S. (1931), "The Analysis of Variance in Cases of Non-Normal Variation," *Biometrika*, **23**, 114–133.

Rousseeuw, P.J. and Leroy, A. (1987), *Robust Regression and Outlier Detection*, New York: Wiley.

Stahel, W.A. and Weisberg, S. (eds.) (1991), *Directions in Robust Statistics*, New York: Springer-Verlag.

Staudte, R.G. and Sheather, S.J. (1990), *Robust Estimation and Testing*, New York: Wiley.

Student (1908), "The Probable Error of a Mean," *Biometrika*, **6**, 1–25.

Tukey, J.W. (1960), "A Survey of Sampling from Contaminated Distributions," in *Contributions to Probability and Statistics, Essays in Honor of Harold Hotelling*, eds. I. Olkin et al., 448–485, Stanford, CA: Stanford University Press.

Welsch, R.E. (1980) "Regression Sensitivity Analysis and Bounded-Influence Estimation," in *Evaluation of Econometric Models*, eds. J. Kmenta and J.B. Ramsey, 153–167, New York: Academic Press.

APPENDIX: Minitab Macros

```
#huber1.mtb
#This Minitab macro calculates the first iteration of Hu-
ber
#regression estimates, based on least squares as starting
values.
#It assumes that y is in c1 and the x variables are in
c2–ck1.
#Up to 9 x variables are allowed.
#It asks the user to enter
#k1=the number of x variables + 1
#k4=constant in the Huber psi function
note Enter the number of x variables + 1 (ie. p + 1)
set 'terminal' c50;
nobs=1.
copy c50 k1
note Enter the constant used in the Huber psi function
set 'terminal' c50;
nobs=1.
copy c50 k4
brief 0
name c11 'dfits',c12 'LSres',c13 'LScoef',c14 'hi'
name c15 'hub',c16 'w',c17 'Adiff',c19 'Wcoef',c29 'Wres'
let k3=k1–1
let k2=count (c1)
regress c1 k3 c2–ck1;
dfits 'dfits';
residuals 'LSres';
coeff 'LScoef';
hi 'hi'.
let k40=median ('LSres')
let c30=abso ('LSres'–k40)
let k40=1.483*median (c30)
let 'hub'=k4*k40/abso ('LSres')
rmin 1 'hub' into 'w'
regr c1 k3 c2–ck1;
weights 'w';
resids 'Wres';
coeff 'Wcoef'.
let 'Adiff'=abso ('LScoef'–'Wcoef')
brief 1
note least squares estimates
print 'LScoef'
note initial set of Huber weights
print 'w'
note initial Huber estimates
print 'Wcoef'
note absolute difference between LS and initial Huber
estimates
print'Adiff'
let k5=max ('Adiff')
note max absolute difference between LS and initial Hu-
ber estimates
print k5
brief 0
let k10=20
let k20=30
brief 1
end
```

```
#huber2.mtb
#This Minitab macro iteratively calculates Huber re-
gression
#estimates. The macro should be executed until the
#estimates have been found to the desired accuracy,
#as measured by k5. Up to 10 iterations are allowed.
#This macro assumes that the macro 'huber1' has been
run.
brief 0
let 'hub'=k4*k40/abso ('Wres')
rmin 1 'hub' into 'w'
regress c1 k3 c2–ck1 c18 c44;
weights 'w';
coeff ck10;
residuals 'Wres'.
let k9=k10–1
let k19=k20–1
brief 1
note latest set of Huber weights
print 'w'
note latest set of Huber estimates
print ck10
let 'Adiff'=abso (ck10–ck9)
print 'Adiff'
let k5=max ('Adiff')
note max absolute difference between last 2 sets of Hu-
ber estimates
print k5
brief 0
let k10=k10+1
let k20=k20+1
end
```

```
#huber3.mtb
#This macro updates the estimate of sigma once after
#convergence of the betahats.
#This macro then calculates an estimate of the variance-
#covariance matrix of the Huber regression estimates.
#This macro assumes that both the macros 'huber1' and
'huber2'
#have been run.
#The residuals and fitted values from the final iteration
#of the Huber regression are stored in
#'FinlWres' and 'FinlWFV', respectively.
brief 0
let k40=median ('Wres')
let c30=abso ('Wres'-k40)
let k40=1.483*median (c30)
let 'hub'=k4*k40/abso ('Wres')
rmin 1 'hub' into 'w'
regress c1 k3 c2-ck1 c18 c44;
weights 'w';
coeff ck10;
residuals 'Wres'.
let k9=k10-1
let k19=k20-1
brief 1
note final set of Huber weights
print 'w'
note final set of Huber estimates
print ck10
let 'Adiff'=abso (ck10-ck9)
print'Adiff'
let k5=max ('Adiff')
note max absolute difference between last 2 sets of Hu-
ber estimates
print k5
brief 0
let k10=k10+1
let k20=k20+1
brief 1
name  c40  'diff',c41  'ind',c42  'wt',c43  'FinlWres',c44
'FinlWFV'
let 'diff'=k4-abso ('Wres'/k40)
let 'ind'=0.5*(sign ('diff')+1)
regress c1 k3 c2-ck1;
weights 'ind';
xpxinv m1.
let k21=k20-1
let 'FinlWres'='Wres'
let 'wt'=('w'**2)*('FinlWres'**2)
regress c1 k3 c2-ck1;
weights 'wt';
xpxinv m2.

inverse m2 into m3
multiply m1 by m3 into m4
multiply m4 by m1 into m5
let k30=k2/(k2-k1)
multiply k30 by m5 into m6
brief 1
note estimated variance-covariance matrix of the Huber
coefficients
print m6
diagonal of m6 into c50
let c50=sqrt(c50)
note estimates of the standard errors of the Huber coef-
ficients
print c50
end
```

Index